The 100 Greatest Inventions of All Time

The 100 Greatest Inventions of All Time

A Ranking Past and Present

Tom Philbin

CITADEL PRESS
Kensington Publishing Corp.
www.kensingtonbooks.com

CITADEL PRESS BOOKS are published by

Kensington Publishing Corp.
850 Third Avenue
New York, NY 10022

Copyright © 2003 Tom Philbin

All rights reserved. No part of this book may be reproduced in any form or by any means without the prior written consent of the publisher, excepting brief quotes used in reviews.

All Kensington titles, imprints, and distributed lines are available at special quantity discounts for bulk purchases for sales promotions, premiums, fund-raising, educational, or institutional use. Special book excerpts or customized printings can also be created to fit specific needs. For details, write or phone the office of the Kensington special sales manager: Kensington Publishing Corp., 850 Third Avenue, New York, NY 10022, attn: Special Sales Department; phone 1-800-221-2647.

CITADEL PRESS and the Citadel logo are Reg. U.S. Pat. & TM Off.

First printing: August 2003

10 9 8 7 6 5 4 3 2 1

Printed in the United States of America

Library of Congress Control Number: 2003101252

ISBN: 0-8065-2403-0

CONTENTS

Introduction *vii*
1. Wheel *1*
2. Light Bulb *5*
3. Printing Press *8*
4. Telephone *10*
5. Television *13*
6. Radio *17*
7. Gunpowder *20*
8. Desktop Computer *23*
9. Telegraph *26*
10. Internal Combustion Engine *29*
11. Pen/Pencil *33*
12. Paper *35*
13. Automobile *39*
14. Airplane *41*
15. Plow *45*
16. Eyeglasses *48*
17. Atomic Reactor *51*
18. Atomic Bomb *54*
19. Colossus Computer *57*
20. Toilet *60*
21. Rifle *62*
22. Pistol *65*
23. Plumbing *68*
24. Iron-to-Steel Process *71*
25. Wire *74*
26. Transistor *77*
27. Steam Engine *79*
28. Sail *82*
29. Bow and Arrow *84*
30. Welding Machine *87*
31. McCormick Reaper *89*
32. Jet Engine *93*
33. Locomotive *95*
34. Anesthesia *98*
35. Battery *101*
36. Nail *104*
37. Screw *107*
38. X-Ray Machine *110*
39. Compass *113*
40. Wooden Ships *115*
41. Stethoscope *118*
42. Skyscraper *120*
43. Elevator *125*
44. Clock *127*
45. Chronometer *130*
46. Microscope *134*
47. Braille *137*
48. Radar *139*
49. Air Conditioning *141*
50. Suspension Bridge *144*
51. Thermometer *147*
52. Incubator *151*
53. CT Scan *153*

54. MRI *156*
55. Drywall *158*
56. Electric Motor *161*
57. Barbed Wire *163*
58. Condom *166*
59. Telescope *169*
60. EKG Machine *172*
61. Pacemaker *175*
62. Kidney Dialysis Machine *178*
63. Camera *181*
64. Global Positioning System *184*
65. Sewing Machine *187*
66. Film *190*
67. Spinning Jenny *193*
68. Brick *196*
69. Motion Picture Camera *199*
70. Dynamite *202*
71. Cannon *205*
72. Balloon Framing *208*
73. Typewriter *211*
74. Diesel Engine *213*
75. Triode Vacuum Tube *217*
76. AC Induction Motor *219*
77. Helicopter *223*
78. Calculator *225*
79. Flashlight *228*
80. Laser *231*
81. Steamboat *234*
82. Fax Machine *236*
83. Tank *238*
84. Rocket *241*
85. Cotton Gin *244*
86. Windmill *246*
87. Submarine *249*
88. Paint *252*
89. Circuit Breaker *254*
90. Washing Machine *256*
91. Threshing Machine *259*
92. Fire Extinguisher *262*
93. Refrigerator *265*
94. Oven *268*
95. Bicycle *271*
96. Tape Recorder *275*
97. Oil Derrick *278*
98. Phonograph *281*
99. Fire Sprinklers *284*
100. Video Recorder *287*

Acknowledgments *290*
Index *291*

INTRODUCTION

This book rounds up and ranks what I view as the one hundred greatest inventions of all time, "greatest" meaning those inventions that have had the most significant impact on humanity throughout history. But what exactly is meant by "impact"? Does it mean preserving and prolonging life, making life easier or better, or changing the way we live? The answer is all of the above and perhaps more—or less—because it is my belief that one cannot apply narrow criteria to an invention's importance but must get an overall sense of it.

In compiling this list, I found it necessary to define, as much as possible, what is meant here by an invention as opposed to a discovery. Originally, for example, I thought penicillin would deserve a place on my list. After all, it was the first antibiotic, saved countless lives (indeed, infection would have decimated the wounded in World War II if penicillin hadn't been available), and led to the development of many other antibiotics. But, in thinking it through, I couldn't characterize penicillin as an invention because it didn't come out of thin air, a product of creative thinking. Instead, it was discovered in 1928 by careful observation: while examining a petri dish containing staphylococcus (a bacterium that causes disease in humans and animals), Sir Alexander Fleming noticed that it had been accidentally contaminated with a mold and that every bacteria that came into contact with the mold had disappeared. Further investigation revealed that the mold could kill a wide variety of bacteria, and penicillin was on its way. (Just prior to World War I, two scientists were able to synthesize it into usable form.)

On the other hand, the telephone resulted because the inventor, Alexander Graham Bell, dreamed of a device that would allow people to speak with one another out of earshot. He worked hard to create such a machine until one day he uttered into the device the words "Mr. Watson, come here, I want to see you."

The essential difference, then, between an invention and a discovery is that one is an act of pure creation, while the other is not. In some cases, truth be known, it will not be that simple to know definitively whether something belongs on the list or how high it should be ranked. For example, does insulin belong? Does plastic? Does language? All have had a huge impact on humanity. But insulin, like penicillin, was discovered. So was plastic. Language, on the other hand, is a human faculty that is rooted in the brain's chemistry and that has evolved over time. In other words, language is the product of evolution, not invention.

One question that surely must be asked in this process will be: What would life be like without the invention? For example, if the blender was not around, what would its impact be? Not much (except for bartenders). On the other hand, consider what life would be like without the telephone, television, radio, airplane, or internal combustion engine. Profoundly different.

I see the entries in this book as, hopefully, going beyond mere information. There will be heavy use of anecdote and interesting detail, the idea being to try to give readers not only information, but an enjoyable reading experience.

Have fun.

The 100 Greatest Inventions
of All Time

1

Roman chariot. *Photofest*

Wheel

Just look around and try to find something in your home that has absolutely nothing to do with a wheel. Almost every machine, every device, every man-made object has a wheel linked to it in some way.

While the exact time and place of the invention of the wheel is anyone's guess, it's widely held that the wheel started out as a rolling log. It is suspected that from there, it became cross-sections of a log, a somewhat heavy and breakable wheel but at least it could roll. Earlier methods of transporting objects already existed like the simple "sledge" or the "travois," which consisted of two poles with a frame slung between them, dragged behind either an animal or the traveler, but this clearly was inferior to the wheel.

What is known for sure is that among the earliest wheels were three planks put together with struts and carved into a circle. This was certainly a stronger way of making a wheel than a solid slab, especially bearing in mind that the wheel predated the invention of roads. The oldest pictograph depicting one of these wheels is Sumerian, dated around 3500 B.C., and shows it mounted on a sledge.

The change in the wheel that made it much lighter and more practical was the spoke, appearing on the scene around 2000 B.C. on chariots in Asia Minor. At this point, the wheel was a mode of transport, with accompanying carts or wagons developing along whatever lines were asked of it. Agriculture, long-distance trade, and warfare all needed wheels.

The chariot in particular evolved from a heavy four-wheeled design, drawn by two to four wild asses called "onagers," to the sleek, horse-drawn two-wheeled kind we are familiar with from films and television. By achieving a combination of a light, sturdy design of the carriage, the lighter axle and spoked wheel assembly, and a good harness for the horses, the chariot revolutionized warfare. The great armies of the second millennium B.C., like those of Egypt, the Hittites of Anatolia, the Aryans of India, and the Mycenaeans of Greece, made use of this speedy, highly maneuverable vehicle of war. The chariot wrought devastation from China, to Minoan Crete, to as far as Great Britain, straight through to the age of Alexander the Great when it was replaced by the cavalry.

A crucial contribution by the Romans to the advancement of the wheel was extensive road building. Building and keeping an empire required good communication and mobilization of resources, trade, and military. Roads allowed that. Their roads lasted for centuries. Indeed, a number of the roads are still used in Great Britain.

As time went by, the wheel itself also continued to improve. Iron hubs were developed and gave the wheel tremendous power at the center where it was mounted on greased axles. Even broken wheels could be rebuilt around the hub, making it an indispensable part of the design. Also, the concept of a "tire" arose in the form of an iron band or ring that was stretched by heat around the wheel rim and then contracted as it cooled, making the wheel not only resilient where it had contact with the ground, but also drew the entire construction into a tighter, more solid unit.

But the wheel, as hinted at earlier, should by no means be looked on merely as an aid to ground transportation. It went from the chariot to the tank, from the wagon to the train and from the cart to the automobile. But certainly, none of these things would have developed without the other uses to which the wheel was put.

Long before the first ostrich was yoked to a racing chariot, though precisely when is impossible to say, the "potter's wheel" became an important step forward in the art of pottery making. No one can say precisely when it was developed, but the first evidence of the potter's wheel is from Mesopotamia around 3500 B.C. A piece of clay, which up to that time had been shaped by molding by hand, could be "thrown" onto the wheel as it was spun and with either just hands or hands and tools, combined with the force of momentum, a symmetrical form could be made. These forms included bowls, pots, and containers of all different kinds. Far more than just decorative in nature, the vessels were the only secure storage for dry goods, beverages, oils, foods, and grains among numerous other uses. Pottery was not only functional for the storage of trade goods, but also an essential trade item in and of itself, as it was traded within and between cultures by ships and wagons.

As important as any other aspect of the wheel, however, is the simple fact that circular or rotary motion had been harnessed and put to whatever task or device to which the human imagination could ascend. Imagine, for instance, the early "waterwheel," consisting of clay pots attached to a large wheel suspended over moving water. The running water would propel the wheel on a horizontal axle as it filled the pots with water. The momentum would bring the full pots to the height of the wheel, where they would pour their contents into a waiting shallow trench or trough that would then track the water to a different location, such as an agricultural field.

Or, the force of the wheel propelled by water, wind, or animal could be an end in itself if the wheel was on a shaft that could move another wheel. Think of the sheer power and dynamics as wind or water turned a great wheel attached to shafts with a massive millstone at the other end, grinding unimaginable amounts of fresh grain, instead of people or animals having to do it.

Indeed, the wheel, with its shaft or axle, became the one invention that led to many others. From the giant Ferris wheel to the barely visible gears of a watch, the wheel was the master of the Industrial Revolution.

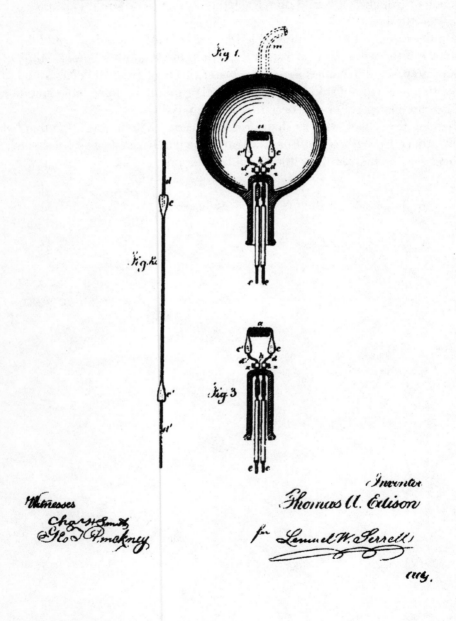

Thomas Edison's 1880 patent drawing. *U.S. Patent Office*

Light Bulb   2

There are a number of myths about Thomas Alva Edison, who is generally and probably accurately thought of as the greatest inventor who ever lived. Many people see him kind of as a Huck Finn character in a rumpled dark suit, hair askew, a harmless old professorial kind of guy. In fact, Edison was a work-obsessed, sometimes ruthless, egoistic man who could be obscene and a little crude. Once, for example, when offered a spittoon, he refused it, saying that he used the floor "because you can't miss."

Another myth was that he invented the incandescent light bulb. There were plenty of people who had created incandescent light bulbs before Edison, some as much as thirty years before him. But none worked that well. His achievement was to invent a bulb that really worked, and in the real world.

Edison's involvement with incandescent lighting started in the spring of 1878, when, at the age of thirty-one, he took a vacation with George Barker, a fellow college professor. During the trip, Barker suggested to Edison, who was already world famous for inventing the phonograph, as well as several other things, that his next goal should be to put electric lighting in America's homes.

Edison was taken with the idea. When he returned to the "invention factory" he had built in Menlo Park, New Jersey, he put together a team of experts and announced to the world that he was going to bring lighting to the American home in six weeks, which proved to be a highly optimistic prediction.

From the beginning, Edison had a vision of creating a bulb that would work in an electrical system where the bulb would require only a small amount of current to operate, last a long time, and where the power was "subdivided." If one bulb went out, none of the others would; also, each bulb could be individually controlled with the flick of a switch. Hence, as he and his team worked to create the bulb, they worked on creating the system as well because neither was good without the other.

The incandescent bulb is a simple device and the core science behind it is the phenomenon of electrical resistance. An electrical current is fed into a material that resists it to varying degrees, thus causing the material to heat up and ultimately glow, providing light.

Pre-Edison incandescent bulbs had suffered from a variety of problems, but two were major. The filaments that the electric current was run through could not stand the heat and burned or melted (if metal). In the open air, this would happen within seconds or minutes, so inventors encapsulated or sealed off the filament in a glass globe and then pumped out the oxygen, creating a vacuum in which the filament could withstand the heat better.

Edison knew that he had to create a super filament, because in order to be practical and use only a little bit of current, it had to be thin, and according to Ohm's law of electrical resistance, this would mean it would be taking a tremendous amount of heat. But this, in turn, would allow the copper conductors supplying power to the bulbs to be significantly smaller. As author Matthew Josephson says in his biography *Edison,* "only one one-hundredth of the weight of copper conductor would be required for such a system as compared with that of the low-resistance system."

In their quest for the perfect filament, Edison and his team tested a number of them and finally settled on platinum, which had a very high melting point, 3,191 degrees F. Simultaneously, others on Edison's team worked on developing better methods for pumping oxygen out of the glass globe to create a better vacuum.

The bulb with the platinum filament worked, but only for about 10 minutes, before melting. Moreover, platinum was a rare and expensive metal, not practical at all. Edison and his team tested many other materials, some sixteen hundred in all and continued to try to make the vacuum inside the globe better and better—and better. But they couldn't find anything that worked well enough.

Then one day, like a detective who has actually held the key to a mystery but had set it aside and comes back for another look, Edison returned to the testing of carbon as a filament, which he had done a year earlier but had cast aside. In the interim, he had solved some problems. A much better vacuum was being created by a "Sprengal pump," which exhausted all but one-millionth of a part of oxygen in the bulb, and he had found a way to eliminate gases that carbon had a propensity to absorb in its porous state and that would hasten its demise.

Edison knew carbon had one great thing going for it. It had a high melting point: 6,233 degrees F, or about 3,500 C. Edison calculated that to work with the proper resistance the filament should be $1/64$ of an inch in diameter and 6 inches long. To make it, he scraped lampblack off gas lamps and mixed this carbon with tar so he could form it into filament shape. Tests made with this filament showed that it would burn for 1 to 2 hours before self-destructing.

But Edison had become convinced that since "tarred lampblack" worked so well, maybe there were other materials that, when turned into carbon, would work even better. With this in mind, he tested an ordinary piece of cotton thread that was turned into carbon by baking it in an earthenware crucible.

The filament was delicate, and a few broke on the way to being installed in the test lamp, but finally the team got the slender reed of material clamped in place, a glass globe enclosed it, the oxygen exhausted, and the current turned on. It was late in the night of October 21, 1879.

The men were used to filaments dying out soon. But this one didn't. It gave off a feeble, reddish glow—giving off about one percent of the light of a modern 100-watt bulb—and thrillingly burned on and on—and on. Eventually, Edison began turning up the current more and more—the bulb getting brighter and brighter—until the filament broke. It had burned for 13½ hours and everyone knew that that feeble little bulb heralded the age of the electric light.

Edison, of course, didn't stop there. He examined the filament under a microscope and realized that the high-resistance carbon he needed must come from materials that were tenacious, fibrous in structure, and, very important, some form of cellulose. Ultimately, Edison used bamboo imported from Japan that burned for 900 hours.

It took Edison only three years—a phenomenally short amount of time—to create and install the electrical system that would make the light practical. His company, the Edison Electric Light Company, constructed a power station at Pearl Street in New York City, running wires through pipes that were once gas pipes to fledgling customers. At first, he had just eighty-five subscribers. There were kinks and bugs in the system, but as these were worked out and the bulbs improved, more and more subscribers came on board. By the turn of the twentieth century, a million people had electric light in their homes. Today, tungsten (the filament) and nitrogen (instead of a vacuum) make up the light bulb.

How important was the light bulb in the scheme of things? One could seemingly speak endlessly about its importance, and when other items on the top ten list, such as gunpowder (which led to both freedom and more people dying) and the internal combustion engine (which essentially put the world on the road and changed the face of commerce) are compared to it, they simply do not have the significance to rise above it. The light bulb changed night, in a sense, to day. People could read, study, stay up later, be operated on, and go out for a late-night dinner and movie. Robert Freidel, coauthor of *Edison's Electric Light,* put the impact of the light bulb very well, "It remade the world in which people worked and played and lived and died.... It was the kind of invention that reshaped the world, and the way people looked at the possibilities in the world."

3

Nineteenth century printing press. *New York Public Library Picture Collection*

Printing Press

Anthropologists credit the advent of writing as the turning point between prehistory and history. Writing enabled thoughts to be recorded. Later, the printing press allowed printers to quickly make multiple copies of pages in books. For the first time in history, the thoughts and ideas of great minds were communicated to masses of people via books, which, until then, were produced in limited quantity for the clergy and nobility and were handwritten in Latin. To state the printing press's impact succinctly, it turned a generally illiterate world literate.

It all started with wooden blocks with raised letters on one side. The blocks were arranged in a specific order inside a frame and inked, and then a sheet of paper was pressed against the blocks. When the paper was removed, an inked copy of the letters remained.

Printing by moveable type enabled one person to do the work of many. In one day, one person could produce what would take a year for a scribe to do.

But there was a problem with wooden blocks. Over time and use, they began to disintegrate and new ones had to be made. Enter Johannes Gutenberg, a German printer. Gutenberg developed a mold of an individual letter out of a metal alloy that lasted over time and could be reused over and over again and remain intact. In fact, his method of mechanical reproduction of printed material was so valid that no significant change was made in printing for another five hundred years.

To put Gutenberg's invention in better perspective, most books at the time were produced by and for the church using the process of wood engraving. This required the craftsman to cut wood to produce templates for both text and illustrations and was extremely time consuming.

As with wooden blocks, when one page was complete, the block would be inked and another sheet of paper was then pressed on it to produce an inked image. Using this process, very few books could be produced in a year, which perhaps wasn't too much of a problem at the time since only church officials and nobility could read.

In 1455, what is known as the Forty-two Line Bible (also known as the Gutenberg Bible) was published in Mainz. It is considered to be the first substantial publication and took Gutenberg two years to complete. His invention allowed the printer to not only make words of the individual letter molds, but also to arrange them in even lines and lock a lot of them together in a single template.

This system allowed printers to do something they had never been able to do before: produce thousands of copies of one page. For the first time, book production speed was increased significantly. A typical press in the fifteenth century could produce five books a year, which may not sound like much but it was at the time.

Gutenberg's invention also served another purpose—it allowed people to read. This set off an explosion of literacy: Philosophical tracts and scientific accomplishments of the day became accessible, allowing people to move beyond the strong religious dogma of the day and to take more of a secular and rational approach to learning and the exploration of the natural world.

By the sixteenth century, the printing press had created a new industry. Typically, a large printing outfit would be made up of five workers. Three would work the press and two served as compositors.

During this period, the work was still tedious and slow. Many times, molds had to be made for the type, although eventually this work was taken over by independent type founders.

As printing spread across Europe, printers eventually found their way to London. Most of them could only survive if they started their businesses in major cities and towns. But in 1563 the Artificers Act was passed in England, which required workers to stay in the parish in which they were born. This act stifled progress in the printing trade because it prevented printers from finding enough people who were interested in working in the trade.

Over time, the most significant development in printing was the creation of different type styles. The most important of these, and one that became very common, was the extended use of Roman in the second half of the sixteenth century. It later became the accepted type that was used in keeping with the qualities of steel, and Roman therefore replaced the earlier Gothic styles throughout most of Europe.

4

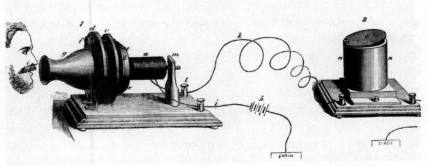

Alexander Graham Bell's telephone. *New York Public Library Picture Collection*

Telephone

The irony of the invention of the telephone was that it was marked by miscommunication, or no communication. Indeed, many inventors worked alone, misunderstood earlier discoveries, or wasted time coming up with results someone else had already achieved.

For instance, Alexander Graham Bell is credited with inventing the telephone. On March 10, 1876, using his new device, he uttered the famous words to his assistant, "Mr. Watson, come here, I want to see you." He was not, however, the only person exploring what would later be known as the telephone. Indeed, he only got credited with inventing it by hours, though others had achieved what he had. For example, Elisha Gray also filed a patent application for a telephone only a few hours after Bell, and if his patent had beaten Bell's I might well be focusing on his achievement instead.

What's more, inventors were not above appropriating the technology of others for their own end. For example, Bell had not built a working telephone, but he did make one work three weeks after filing his patent application by using Gray's "Notice of Invention."

Although both men were tenacious, hard working, and inventive, it was

Bell's knowledge of acoustics (the study of sound) that gave him the edge over Gray. Bell did know a little about electricity, a necessary ingredient for a telephone to work, but he knew a lot about acoustics. Indeed, like Gray, many other inventors working on similar projects knew more about electricity than acoustics, which also translated into an inability to bring the two disciplines together to create the telephone.

Some historians believe the earliest reference to such a device was from Francis Bacon in 1627 in his book *New Utopia*. Here, he referred to a long speaking tube that was conceptual only; a telephone cannot work without electricity, and Bacon made no mention of that. The electrical ideas needed were known by the early 1830s, but it wasn't until 1854 that anyone even suggested transmitting speech electrically.

The first real spat in the telephone's evolution occurred in 1729 when an English chemist named Stephen Gray transmitted electricity over 300 feet of wire. Later, in 1746 two Dutchmen developed a "Leyden jar" for storing static electricity. It acted as a battery to store energy, but the big drawback was that it stored such a small amount of electricity that it couldn't be used for any practical purpose. But it was the start of something bigger.

Static electricity could be seen and could make hairs on the back of one's neck stand up. Later, in 1753 an anonymous writer suggested that electricity might transmit messages. His experiments required wires by the dozen and an electrostatic generator to electrify the line that would attract paper on which letters were printed by static charge at the other end. By noting which letters were attracted, one might spell out a message. Although the crude system worked, it was, of course, severely limited and required a ganglion of wires.

It wasn't until the battery was invented that experimentation on the telephone reached a new level. The battery did something the electrostatic generator could not—it provided sustained, low-powered electric current. It was chemically based and although it couldn't produce enough electricity to work machinery, with improvements it would over time.

Although electricity was now available, it amounted to just one half of the equation for telephones. Transmitting speech required an understanding of magnetism.

Enter Danish physicist Christian Oersted in 1820. In his famous experiment at his University of Copenhagen classroom, he pushed a compass under a live electric wire. This caused its needle to start turning north, as if attracted by a large magnet. He had discovered something startling: electric current creates a magnetic field.

Then, in 1821 inventor Michael Faraday reversed Oersted's experiment and discovered electrical induction. He was able to get a weak electrical current to flow in a wire revolving around a permanent magnet. In other words, a magnetic field caused an electric current to flow in a nearby wire.

The result was stunning. Mechanical energy could now be converted to electrical energy. This result led, years later, to such mechanical things as turbines through which water flowed and burning coal to produce electricity.

From the humble original telephone—a crude device consisting of a wooden stand, a funnel, a cup of acid, and some copper wire—to the modern-day telephone, the device works pretty much the same. In modern electrical transmitters, a thin plastic sheet (much like a human eardrum—the tympanum—that works by the same principle) is coated on one side with a conductive metallic coating. The plastic separates that coating from another metallic electrode and maintains an electric field between them. Vibrations caused by speech produce fluctuations in the electrical field, which in turn produce small variations in voltage. The voltages are amplified for transmission over the telephone line.

Simply put, the modern-day telephone is an electrical instrument that carries and varies electric current between two mechanical diaphragms. It duplicates original sound from one diaphragm and transfers it to the other. Simple, but at the same time profound in its impact.

Television   5

Most people would think that the development of television (TV) followed the perfection and popularization of the telephone, motion pictures, and radio, but in fact, early research and experimentation began during the mid-nineteenth century! Theoretical proof of the relationship between light and electricity—essential to TV transmission—was detailed by Michael Faraday during a series of experiments in the 1830s, and there were other insights as well.

Although these developments would seem to pave the way for an early birth of TV, there were numerous technical obstacles in the way, including the fact that wireless transmission of sound—an essential for TV transmission—was unknown.

TV signals are transmitted electronically, so light waves, which must be sent, must be converted into electronic signals, and this could not be done simply. Light waves, however, are infinitesimally small and cannot be converted directly into electronic signals by a simple mechanical coupling. In addition, picture information is far more complex and reaches far higher frequencies than do sound waves. The first breakthrough in solving this dilemma came in 1873 when it was discovered that the element selenium showed varying electrical resistance in proportion to the amount of light focused on it. It was now possible to convert light into an electronic signal or "pulse" that, in theory, could be sent along a cable or transmitted through the air.

In 1883, German engineer Paul Nipkow introduced a device utilizing a rotating scanning disk that was perforated with small holes in a spiral pattern. This disk broke down a picture into a series of dots that was in turn focused on a photocell. The photocell sent a series of electrical pulses to a receiver, where another scanning disk was placed in front of a light and "unscrambled" the dot pattern into a picture. It was a crude and fuzzy picture, but it was the first real TV picture. The Nipkow mechanical scanning system started a parade of imitations and improvements during the next quarter of a century. By 1925, Charles Francis Jenkins—using a mechanical scanning system—was actually sending signals out over the "air" from his laboratory in Washington, D.C. In Europe, the Scottish inventor John Logie Baird presented a public demonstration of his mechanical TV system in 1926.

It was realized early on in the late nineteenth and early twentieth centuries that mechanical TV transmission could never be made satisfactory. But with the

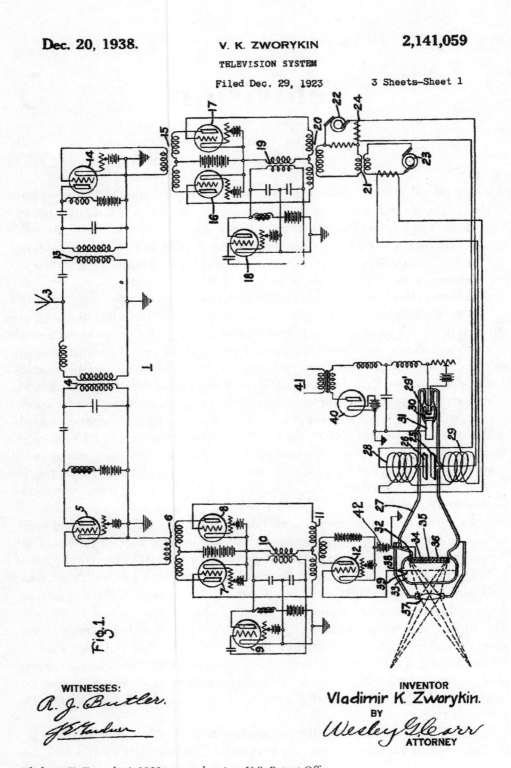

Vladimir K. Zworykin's 1923 patent drawing. *U.S. Patent Office*

rapid strides made in the fields of radio, X-rays, and physics, the problems were soon to be resolved. Radio, of course, had become a reality in the 1900s with Guglielmo Marconi's wireless transmissions and the development of the vacuum tube by John Fleming and Lee De Forest. The cathode-ray tube—used to produce medical X-rays—was another essential element of TV technology.

In 1906, Karl Braun found that when he introduced a magnetic field into a cathode-ray tube he could vary the course of its electron stream. A year later, Alan Campbell Swinton advocated such a cathode-ray tube as a receiving device for picture images. Russian scientist Boris Rosing soon developed and patented such a cathode-ray device. Studying under Rosing, Vladimir K. Zworykin worked on combining vacuum tube technology and the new cathode-ray receiver into a practical TV system.

Shortly after World War I ended, Zworykin immigrated to the United States, a move that was to herald the final perfection of modern TV. Zworykin envisioned and built what he named the "iconoscope," a transmission tube that used electrons to scan a picture and break it down into a series of electronic signals. The picture was focused on a plate—the "mosaic"—that was coated with microscopic globules of a light-sensitive material. When an electron beam was aimed at the plate, a current was sensed in relation to the quantity of light falling on the plate.

The Russian scientist then worked on the receiving device, which he dubbed the "kinescope," adapting the Greek word *kinema* (movement). Incidentally, the word "television" is a combination of the Greek word *tele* (at a distance) and the Latin term *video* (to see). Zwyorkin's receiving tube—which essentially reverses the process of the iconoscope—was combined with the latter and demonstrated publicly in 1929.

During the same time period, Philo T. Farnsworth, a young experimenter from Idaho, created a TV system closely paralleling Zwyorkin's. Farnsworth's "image dissector" was basically similar to the iconoscope, but passed the electron stream through a small aperture before transmission. Farnsworth later received many patents on TV technology and with the Philco Corporation became an early manufacturer of TV receivers.

Zwyorkin, in the meantime, began working for broadcasting mogul David Sarnoff at the Radio Corporation of America. Sarnoff was one of the first in the business community to see the potential of TV.

A landmark year in TV progress was 1939. A regular schedule of broadcasts was started by NBC in that year, seen over about a thousand receivers installed in hotels, taverns, and store windows. In 1940, the first "network" telecast took place when a program originating at NBC in New York City was retransmitted by a station in Schenectady, New York.

Although much TV technology was adapted to the war effort in the form of radar and other detection devices, the medium itself played no major role in World War II. But by the end of the war, Sarnoff—now a brigadier general—and other TV entrepreneurs were anxious to pick up where they had left off in 1941.

They would be helped by a major technological development by Zwyorkin and his staff.

Although Zwyorkin's iconoscope had made TV practical, it did not make it easy or perfect. The "ike"—as it was popularly referred to by TV personnel—produced clean, sharp pictures, but was not very sensitive to light. In bright sunlight, all was well, but in the studio vast amounts of light—much more than were needed by the motion picture industry—were required. Heat levels reached an excess of 100 degrees F and actors and actresses needed makeup, including dense eye shadow and lipstick, to compensate for the glare shed by the old electric arc lamps. Zwyorkin and his staff to the rescue!

They first used "secondary emission" gain to increase sensitivity by about 10 percent. They also perfected a "low-velocity" scanning circuit and other innovations that resulted in the new "image orthicon," a tube that could sense images illuminated by candlelight! This new tube was ready by 1945 and became the standard black-and-white camera tube.

WNBC went back on the air in late 1945 and broadcast films of Japan's surrender. Others, such as CBS and the new Dumont network, also resumed regular broadcasting by late 1945 and early 1946. By 1948, thirty-six stations were on the air nationwide with about one million sets installed in homes and public places.

Color TV became reality by 1953, with the first sets put on sale in 1954. They were $1,000 apiece for a small picture, but by the turn of the twenty-first century, color TV was no big matter.

TV technology continues to advance along with its connections to new media technologies such as DVDs and the Internet. But there's just no telling how much more important television will be in our lives and the lives of future generations.

Guglielmo Marconi, inventor of the radio. *New York Public Library Picture Collection*

Radio

Like a number of other inventions, radio was dependent for its birth on two others: the telegraph and the telephone. And like other inventions, there were a number of people involved.

At the center of radio's birth is Guglielmo Marconi, an Italian physicist who essentially took the ideas of others and assembled them into the first successful "radiotelegraph." Before Marconi, there was James Maxwell, a Scottish physicist who first postulated in the 1860s that it was possible to send electromagnetic radiations through what was known as the "ether." Heinrich Hertz, another physicist, was able to demonstrate, about twenty years after Maxwell, that such radiations do exist and dubbed them "Hertzian waves." Then, in 1894 Sir Oliver Lodge, an English scientist, sent a signal not unlike Morse code for more than half a mile. Unhappily for them, neither Hertz nor Lodge thought of radio waves as more than a scientific oddity; not something having practical implications.

Of course, they were not joined by everyone. A Russian scientist named Alexander Stepanovich Popov could envision practical applications, including sending and receiving signals from points miles apart, a possible boon when communicating with ships. Indeed, in Russia Popov is hailed as the person who was the true inventor of radio.

In 1895, Popov built a receiver that detected electromagnetic waves in the atmosphere and stated that the receiver might one day be able to pick up generated signals. In 1896, he demonstrated that this could be done in an experiment conducted at what was then the University of St. Petersburg.

As Popov worked in Russia, Marconi worked in Italy. In fact, he conducted a series of experiments at his family's estate in Bologna, one of which was an attempt to boost the power of a signal to send it to a point on the other side of a hill. This he did by securing one end of his transmitter to one end of a long wire that, in turn, ran to the top of a pole.

Despite his early success, the Italian authorities were not interested in his work, so Marconi moved to London. There, he continued his experimentation by narrowing and strengthening the radio beam he was trying to send. With the assistance of a cousin from Ireland, he prepared and received a patent for his device. Recognizing the possibilities, the British Post Office exhorted him to develop it further.

The invention gradually evolved and got stronger and stronger, until at one point Marconi was able to beam a radio signal 9 miles across the English Channel. Emboldened by this, he and his cousin founded the Wireless Telegraph and Signal Company. In 1899, he set up a wireless "radiotelegraphy" station in England to communicate with a station in France that was 31 miles across the channel. Regardless, some scientists said that going further and indirectly was impossible. In 1901, current scientific theory held that it would be impossible to send a radio signal very far across the earth because it was round. Like light, it would travel in a straight line; therefore, it would not be able to follow its curvature.

On December 11 of that year, Marconi set up a test to send a signal some 2,000 miles, from Poldhu, Cornwall, to St. John's, Newfoundland. He signaled the letter *s*, and it arrived, and the world took notice.

There was some mystery involved in exactly how he did it. For the test, he replaced the wire receiver normally used with a device called a "coherer," a tube filled with iron filings that was able to conduct radio waves. No one at the time could explain why it worked, but many believed that it had something to do with the "ionosphere," which reflected electromagnetic rays. In 1924, however, the real reason was discovered: There was an electrified layer in the upper atmosphere capable of reflecting such radiation, which could bounce off this layer and reach its destination.

Following his scientific success, Marconi devoted himself to enhancing his business interests. In 1909, he shared the Nobel Prize in physics with German physicist Karl Braun, a radio pioneer but one better remembered for the cathode-ray oscilloscope, a key component in television.

Marconi's Nobel states that his invention was being used by the principal warships of the Indian and British navies and 298 British merchant ships. A variety of spectacular events continued to advertise radio to the world, not the least of which was the role it played in the capture of the notorious and murderous Hawley H. Crippen and his mistress after the captain of the ship they were on was alerted of their presence via radio. Its importance was also dramatically demonstrated when the *Titanic* sank in 1912.

Radio became, of course, one of the most important inventions ever, and it certainly deserves its place in the top ten of the one hundred greatest inventions list.

7

Battle of Concord Bridge, Revolutionary War. *Photofest*

Gunpowder

Around A.D. 900, Chinese alchemists got quite a surprise when a mixture of potassium nitrate, charcoal, and sulfur was introduced to a flame. The result was a horrible smell, a terrible noise, a cloud of white smoke, and a powerful, quick expansion of hot gases. It was soon discovered that if the powder was ignited in a container, these gases could propel an object with considerable force out of the container's opening and hurl it some distance. The Chinese put this discovery to use in the form of "fireworks" and as signaling devices.

But it took the European mind to bring out the lethal potential of gunpowder, in the form of siege cannons and bombs hurled by mechanical means, though it wasn't done overnight, as it took centuries to perfect it. Indeed, it didn't come onto the European scene until the thirteenth century.

Experimenting with it could, of course, be problematic. While black powder was relatively safe, the basic hand-ground, dry mixture could be hazardous to use. This dry mix, called "serpentine," had a range of virtually unpredictable reactions from a slow fizzle to a spontaneous detonation. What's more, during transport the mix would separate and settle by the density of the ingredients: sulphur on the bottom, then potassium nitrate and charcoal, the lightest material, on top. This required remixing in the field, which was a dangerous proposition, causing clouds of poisonous and possibly explosive dust.

By the 1400s, the basic ingredients were in place, but had not yet been put together in the ratio that would be productive as an explosive and for weapons. Different mixtures and materials were tried. Without the benefit of modern science, the developments were mostly based on the observations of gunners in the field. But some of those gunners, in their empirical wisdom, were amazingly accurate, though their theories, used in practice for years, were not scientifically proven for centuries.

One theory was that larger grains burned more slowly and could therefore provide a longer burning reaction. This was correct because, chemically, gunpowder is a surface-burning agent. The larger the surface area of a grain of powder, the longer it burned. This was particularly useful to know for the cannon, where the objective was to hurl a large projectile as far as possible without blowing apart the cannon. Therefore, the larger the grains of combusting gunpowder, the more slowly the gases would be released. Building this pressure up gradually behind the projectile optimized the cannon's ability to send the projectile flying.

Once proper ratios were established, little changed about gunpowder except the way it was made. Initially ground by hand with mortar and pestle, water-driven grinding stones were introduced. Grinding the ingredients into a watery mud not only helped to make the process less explosive, but also gave it more uniformity and stability. The mud, or "slurry," was subsequently dried in a sheet and then broken apart by mechanical hammerlike "stamps" into granules of various sizes. These granules were then tumbled smooth and sieved to separate them by size ranging from a powder to corn-sized kernels. The finest pieces, those too small to be used, were thrown back in the slurry to be reused.

The process allowed the user to select the grain sizes that worked best. The very largest grains were ideal, as already mentioned, for producing the profound buildup of gasses needed to push a cannonball out the muzzle. Medium grains were best for midrange-sized weapons like hand cannons and muskets, and fine powder was for pistols, which had the lightest, smallest projectiles and the shortest range.

This all had to be done very carefully. As gunpowder was better understood and manufactured, it got to be very powerful, and if the strength used was too great, it could blow apart the weapon.

The ingredients continued to be improved. Advances in the making of pure charcoal were crucial as well as the resulting discovery that different types of wood used for charcoal produced different amounts of gasses and therefore were suited to different purposes. Willow charcoal, for instance, produces less gas per unit than fir or chestnut and considerably less than dogwood. Hence, willow charcoal was particularly suited to the desired slow-gas buildup of a cannon as opposed to the rapid buildup of dogwood, which was superior for small arms.

Ultimately, by the time refinements were perfected and subjected to thorough scientific evaluation, gunpowder was being replaced by nitrocellulose-based propellants, more commonly known as "smokeless powder" or "guncotton." Aside from the obvious advantage of the user's location not being revealed by a large plume of white smoke that gunpowder produced, these plant-based propellants were much more stable in storage and provided an improved control of the rate of combustion. Basically, despite modern refinements, gunpowder could never evolve beyond the fact that it produced 40 percent gas propellant and 60 percent solid discharge that wore out gun bores and made them nearly impossible to keep clean and, therefore, prone to catching fire.

Gunpowder certainly led humanity to more modern methods of destruction, and it sparked—no pun intended—interest in chemistry. But the irony is that today it has gone full circle, mostly being used as the Chinese used it: for signals and fireworks.

8

Late twentieth century desktop computer. *Author*

Desktop Computer

The modern-day desktop computer is often associated with the adjectives used in marketing the machines—"sleek," "smart," and "high-tech"—and is operated by geeks and geniuses like Bill Gates from Microsoft and Steve Jobs from Apple. Unlike a simple invention that came about overnight, the modern desktop computer is simply the latest step of an evolutionary, gradual development. Most people associate the oldest computers with industrial-looking machines that filled a room and computed problems slowly in the 1940s and 1950s. Over time, they gradually grew smaller and speedier in their calculations until the desktop appeared in the early 1980s.

There are two basic kinds of computers. The first is the analog computer. Analog computers perform calculations based on continuously varying quantities such as temperature, speed, and weight. Instead of counting, analog computers "compute" one thing by measuring another.

The first modern computer is credited to Vannevar Bush, an electrical engineer at the Massachusetts Institute of Technology in the 1930s. The computer was what he and his team needed: an answer to the question of how to reduce the time-consuming task of solving mathematical equations required to in turn solve engineering problems. Automating the problem-solving process was what they were after. Finally, in 1936 they unveiled the "differential analyzer."

It weighed 100 tons, had 150 motors, and hundreds of miles of wires. That's a lot of work and equipment for what it accomplished. It was estimated to be a hundred times faster than a human operator using a desk calculator. Although highly successful for their time, by the 1950s many of the most complex tasks performed by analog computers were being completed by faster and more accurate digital computers. Regardless, analog computers are still used today for scientific calculation and spacecraft navigation, among other things.

The second kind is the aforementioned digital computer. This computer is programmable and processes numbers and words accurately and at an enormous speed. It should be noted that the digital computer evolved for the same reason the analog computer did: the never-ending search for labor-saving devices. Although calculating machines have been recorded as far back as the abacus of the fifth century B.C. and the pebbles that merchants in Rome used for counting, none of these early devices was automatic.

It took the Industrial Revolution in the early 1800s for the need to arise for a fast calculating machine that was error free. This was because the revolution in technology was beginning to automate tasks that centuries before had been completed by people. People were generally too slow and sometimes made mistakes.

A person who didn't like mistakes was Charles Babbage, a brilliant young English mathematician. In 1822, Babbage produced a small model of his "difference engine." The machine compiled and printed mathematical tables as a user turned a handle on top of the machine.

The machine was never mass produced, but by that time Babbage had already completed his "analytical engine," an automated, programmable machine that performed many kinds of arithmetic functions. Twenty years later, that technology helped the U.S. government complete census data. The evolution of the digital computer is inextricably linked with World War II and it was this world event that saw the digital computer, coupled with the brilliance of its users, turn the tide of the war. The Colossus was a special purpose computer built by the British to break German codes.

The first "programmable" calculator to be widely known appeared in January 1943 and was 51 feet long, weighed 5 tons, and had 750,000 parts. The machine, known as the Harvard Mark I, was developed by Howard H. Aiken and his team at Harvard University with financial backing by IBM. It could perform addition and multiplication tasks that, according to today's standards, were relatively slow.

The most important characteristic in what people could truly call a computer was a true stored-program capability. The first full-scale operational stored-program computer was unveiled at Cambridge University in May 1949.

The first U.S. commercial computer appeared in March 1951 and had 1,000 12-digit words in its memory and could do 8,333 additions or 555 multiplications per second. It contained 5,000 tubes and covered 200 square feet of floor space, considerably less than only a few years before. The U.S. Census Bureau was the first customer to buy the computer. The first IBM computers entered production at Poughkeepsie, New York. The first order was delivered in March 1953. A total of nineteen computers were sold together, with each capable of doing twenty-two hundred multiplications per second.

The rest, as the saying goes, is history. Today's desktop computers are faster, smaller, have more memory, and can perform many more functions than their predecessors—this thanks to the invention of the microchip.

Computers now play a role in every facet of our lives and they will play ever increasing roles in the way we live and interact. That interactive potential has evolved with growth of the Internet, where people and their computers connect with other people and their computers all over the world.

9

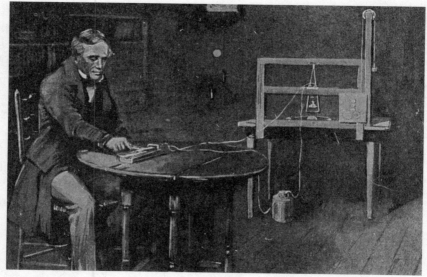

Samuel F. B. Morse, inventor of the telegraph. *New York Public Library Picture Collection*

Telegraph

Interestingly enough, Samuel F. B. Morse, the person who invented the telegraph, started out in life as an artist, specifically a portrait painter. Usually, people who are creative in the liberal arts do not normally get involved in things mechanical, though there are certainly exceptions to this "rule." Indeed, the prime example would be Leonardo da Vinci.

Graduating from Yale University in 1810, Morse set sail for England, where he was to study art. This he did, returning in 1813 and gradually developing into one of America's best portrait painters. He painted various famous personages of the day, one of whom was another inventor, Eli Whitney, who invented the cotton gin.

Morse always had an interest in science. One day in 1832, returning on a ship from another European trip, he overheard something being talked about that stimulated his imagination. It was about the invention of the electromagnet by Joseph Henry, a device, Morse learned, that was capable of sending an electrical impulse along a wire. Indeed, Morse found out that in 1831 Henry had sent an impulse along a mile-long length of wire. An electrical impulse, generated by a battery, traveled along the wire and announced its arrival by ringing a bell secured to a magnetic armature at the other end of the wire.

Morse's idea was to create a communication system using the electrical impulses as his basic language. Seized by the idea, he started to create various magnetic transmitters and receivers, and three years after he had heard the conversation on the boat, he was ready to test prototypes. Further gripped with his mechanical creations, in 1837 he abandoned art altogether and a year later developed the series of electrical dots and dashes that were to become known as "Morse code."

The problem then became for Morse to test his invention on a grand scale, and for this he worked hard persuading the U.S. Congress to fund his project. He did not succeed in convincing Congress immediately, but eventually, it agreed, and a line was run 37 miles between Baltimore and Washington. With bated breath, onlookers watched an operator tap out the message that was received on the other end: "What hath God wrought?"

Though the test was successful, getting the telegraph to be accepted by the public was not a simple task. Many people—spooked by the idea of electrical currents running across the land and worried about their own safety—opposed it.

Morse himself had his own patent problems—as many inventors do—and he was sued and sued by many people who sought to retain the rights to his patents. Eventually, the definitive suit found its way to the U.S. Supreme Court. In 1854, the Court decided in Morse's favor.

Ironically, the one man who did not sue him was the one person who most properly had a claim against him: Joseph Henry. It was Henry who invented the crucial system of relays that allowed the telegraphic signal to be boosted so it could find its destination, but Morse never acknowledged this. Indeed, like some other inventors, he never acknowledged any help from anyone.

Eventually, the problem of the public's acceptance resolved itself. The public came to accept the telegraph, which played a key role in the development of the American West as well as other inventions dependent on electricity.

The telegraph was not without problems, however. Originally, the dots and dashes were sent by an operator varying the time he or she depressed the sending key. But many operators found making these times precise difficult to do. To solve this problem, Morse invented a device that essentially consisted of metal strips mounted on a nonconductive plate that was connected to a metal plate under the other plate. All the operator had to do was move the rod against the metal plate, which automatically resulted in the proper length to produce the dot or dash as required.

As the years went by, the receiving mechanism was also redesigned. First, there was a continuous roll of paper and a stylus that punched out the code, and then an inking machine replaced this. By the mid-1850s, it was discovered that operators were able to write down the code if a certain "sounder" was used. This was generally adopted and its sound became famous in many a movie scene where life was on the line, as it were.

Morse died at the age of eighty-one in 1872. The invention had made him rich, and he became a philanthropist, contributing to temperance and missionary organizations as well as schools.

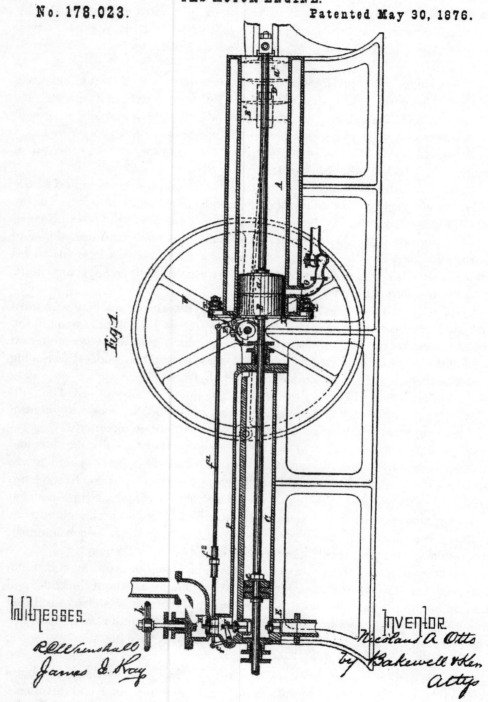

Patent drawing, 1876, by Nikolaus Otto. *U.S. Patent Office*

Internal Combustion Engine 10

Many inventions and developments are double-edged swords, both benefiting and threatening humanity. The internal combustion engine must be placed in this category. It has created pollution and accelerated global warming. But without it, people might not have taken to the air and the highways, farmers and laborers would not have enjoyed shorter and easier working days, and the wide availability of electric power for lighting and home appliances would have taken decades longer to achieve. The internal combustion engine was the "on-board" power for the progress of the twentieth century and is still in our service today.

Steam engines, which used water that was heated by wood and later coal fires, evolved slowly throughout the first millennium. They had serious drawbacks, however, that limited their application. Steam engines were large and bulky. They couldn't be started and stopped quickly and easily. And—this is a big "and"—they were dangerous, with boiler explosions and steam burns happening all too frequently.

The internal combustion engine resolved these handicaps. In the internal combustion process, a piston moves in a cylinder into which an air and fuel mixture is compressed and ignited. The explosion forces the piston to move, thereby creating the mechanical power. External boilers, safety valves, long belts, and linkages are therefore eliminated. The expansion of gases is largely contained, resulting in greater efficiency than steam generation. Therefore, engines of fairly large horsepower, such as 10 to 100, could be built weighing less than a quarter ton. This was to become crucial when lightweight engines were needed to power the automobile and the airplane.

Experimentation with internal combustion principles began long before the "age of steam." Jean de Hauteville used the escaping gases from ignited gunpowder to operate a small, but impractical engine. Famed Dutch engineer Christiaan Huygens and Denis Papin, another Frenchman, also conducted experiments with gunpowder engines in the last decade of the seventeenth century.

A century was to pass before the engine was taken up for its eventual development and practical application. By the 1790s, other possible fuels—explosive gases, alcohol, and soon petroleum distillates—were available in lieu of gunpowder. In 1794, Robert Street was granted a British patent for what can be termed the first true, practical internal combustion engine. It consisted of a cylinder with a piston linked to a pivoting arm that operated a simple water pump. The cylinder—surrounded by a cooling water jacket—extended into a furnace that heated it to a temperature that ignited an air and liquid fuel mixture. The fuel was dripped into the cylinder by gravity and the air had to be hand-pumped while the engine was running, but it did work. Soon thereafter, inventors and engineers set out to improve the Street design.

Proposals were soon being made to compress the cylinder space above the piston before the fuel was ignited, thus increasing power on the "downstroke," and to use hydrogen and air mixtures as fuel. In 1823, Samuel Brown began building and marketing gas-burning engines in England. In 1824, French engineer Nicolas Carnot published the treatise "Reflections on the Motive Power of Heat," which embodied a large part of what was to become the basic theory of modern internal combustion design. Carnot, however, was a theorist and did not actually try to build engines.

Important progress was made by William Barnett in the late 1830s. He put into use the compression principle, proposed in the 1800s, and patented such an engine in 1838. Barnett also constructed the first practical "two-stroke" cycle engine using an external air and fuel pump. The two-stroke design combined the intake/ignition and power/exhaust cycles of the "four-stroke" engine and found wide application in the later development of the diesel and small "utility" gasoline engines. In addition, Barnett—a much unheralded pioneer—devised a "pilot flame" ignition system. This became a popular method for igniting fuel until the electric spark plug was invented.

Inventors labored and tweaked with designs based on the previously described engines throughout the 1840s and 1850s. In 1860, Frenchman Etienne Lenoir successfully built and sold an engine that combined some elements of steam engine technology—it used sliding sleeve valves for intake and exhaust—with illuminating gas as a fuel. Though the engine wasted much of its fuel and was not very powerful, several hundred were sold.

Important practical theory was also disseminated in the 1860s with the work of Alphonse Beau de Rochas. Beau de Rochas outlined several qualifications for improvement of internal combustion engines in a paper published in 1862. He noted that optimum power and efficiency would depend on maximum cylinder volume with minimum cooling surface, maximum rapidity and ratio of the exploding gases, and maximum pressure (compression) of the ignited fuel. He also detailed what was to be the standard progression of the four-stroke operation: intake, compression, power, and exhaust. Beau de Rochas, like the aforementioned Carnot, was strictly a theorist and not a builder. Nikolaus Otto, however, was a builder and he took the Beau de Rochas principles into manufacturing and sold the first modern internal combustion engine.

Otto began building engines in 1867 with the firm Otto and Langen in Germany. His early products were variations on a "free-piston" design taken from the steam engine. This engine made use of electrical ignition and a rack and pinion transmission system; it was noisy and underpowered but was an improvement over the Lenoir-type engine.

In 1876, Otto improved the design of his earlier engines and produced a four-stroke engine, in essence the design still widely used today. Otto was granted a U.S. patent in 1877 and began marketing his engines in the U.S. the following year. Otto and Langen sold 50,000 engines with a combined horsepower of 200,000 by the early 1890s.

Other "parallel" developments took place during the same period, but their full application was only reached during the twentieth century. In 1873, George Brayton invented a two-piston engine that employed constant pressure combustion, a precursor of the gas turbine. In 1895, Rudolph Diesel began work on a "compression ignition" engine in which the heat of the air compressed in the cylinder ignited the fuel without the use of spark plugs.

The years before and after the turn of the twentieth century brought the internal combustion engine into ever-growing application that soon equaled and then eclipsed steam. Charles Duryea adopted the gasoline engine into the new "horseless carriage," while the Wright brothers were first to fly, using a specially designed lightweight gasoline engine. Farmers soon retired their mules and horses and climbed behind the wheel of a John Deere or other brand tractor. Henry Ford put America on wheels driven by an internal combustion engine.

Cylinders have been added—two, four, six, eight, and more—and we've seen devices to limit pollution and save fuel, but the internal combustion engine is still remarkably similar to the Otto pattern of the 1870s. We have put atomic energy to use in large-scale applications, such as electric power generation, but the internal combustion engine has yet to be replaced by electric or other, as of yet, unknown power. For good or bad, it's here to stay, for at least a while.

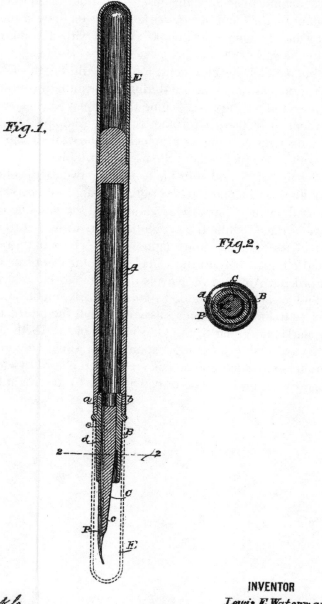

Patent drawing, 1884, by Lewis E. Waterman. *U.S. Patent Office*

Pen/Pencil   11

Nobody can be certain when the first writing instrument was invented, but it's safe to say that it has been around for a long time. For this reason, a discovery in Borrowdale, England, in 1564 is considered the birthplace of the modern-day pencil. As the story goes, an unknown passerby found bits of shiny, black stuff clinging to the roots of a fallen tree—and the material could be used to write and draw. The discovery caused quite a stir and the stuff, graphite (a form of carbon), became known as "blacklead."

Making graphite truly practical for use was a problem because it is naturally soft and brittle. It needed a holder. At first, sticks of graphite were wrapped in string. Later, graphite was inserted into wooden sticks that were hollowed out by hand. Though laborious, the method proved to be productive and the wood-cased pencil was born.

The first patented process used for making pencils was introduced in 1795 by French chemist Nicolas Conté. His patent called for using a mixture of graphite and clay that was fired before it was put in a wooden case. The earliest pencils made this way were cylindrical with a slot. After the clay-graphite mixture (called lead) was inserted into the slot, a thin strip of wood was put back into place.

An important element of Conté's process was his ability to produce a variety of hard or soft leads, depending on how he kiln-fired the powdered graphite. This was important to artists, writers, and draftsmen.

Although the first mass-produced pencils were made in Europe and shipped overseas to the United States, war in Europe cut off imports and America was left to devise its own pencils. Thus, William Monroe, a cabinetmaker in Concord, Massachusetts, made the first American wood pencils in 1812. Apparently, he did something right. He had picked up on earlier pioneers who successfully marketed the instruments, even when they were only available overseas. For example, Benjamin Franklin advertised pencils for sale in his *Pennsylvania Gazette* in 1729 and George Washington surveyed the Ohio Territory with a pencil in 1762.

When pencils were first mass-produced, they were not painted in order to show off the quality wood casing. The early pencils were made with eastern red cedar, a strong, splinter resistant wood that grew in the southeastern part of the United States and eastern Tennessee.

Millions of pencils are made yearly in modern times. They come in nearly every conceivable color and hardness or softness and are designed as specialty pencils that can write on nearly any surface for any reason. It's certainly a tool that is indispensable to tradespeople, artists, and writers.

The pen has an equally interesting history. The first pen-and-paper system hails from ancient Egypt. The scribes of the pharaohs and high priests used reeds with the ends chewed into filaments capable of holding ink.

Over time, as pigments improved, pens evolved into sharpened instruments with slits cut into the ends of them. In the sixteenth century, feather quills were developed and represented a great leap in the quality of writing instruments. They were able to be sharpened, were pliable, and broke less often under the user's hand pressure.

Three hundred years later, in the mid-nineteenth century, metals were used (the fountain pen was developed) but users still had to dip the instrument into an inkwell and write until the tip was dry. In essence, pens in the mid-nineteenth century were used in the same way as reeds had been thousands of years before.

As with many inventions, someone saw a problem with the status quo and wanted to solve that problem. Such was the case in 1884 for insurance broker Lewis Waterman. He was fed up with the inconvenience of dipping the pen into an ink well. Prior to this, the reason that ink reservoirs were not incorporated into pens was because it was difficult to control the ink flow.

Waterman created a solution. To keep pressure on ink flow, he came to the conclusion that air should replace the ink as it was used. To do this, he created two or three channels that permitted air and ink to move simultaneously.

Later, roller balls and ballpoints were developed. The difference between fountain pens and ballpoints is quite extreme. In a ballpoint pen, the ink is directed toward the pull of gravity—that is, pointed toward the paper when held with the point down (in writing). The ink dries immediately and the action is similar to painting a wall with a roller. A roller ball pen differs, still. The first difference is a cap that is required so the ink does not dry out. The second is that the ball doesn't apply ink. Instead, it acts like a flow regulator and friction reducer. Also, the ink is not as liquid as that in a fountain pen.

So far, no one seems to have solved a nagging problem with ball-points: leakage. Hopefully, that will be the next technological breakthrough!

Paper factory. *U.S. Gypsum*

Paper

Just think of what the world would be without paper in it and you can get a sense of just how important it is, and the kind of impact it has had on humanity.

The desire to communicate, of course, came before the means to do it. People started by using clay tablets, silk, bronze, waxed boards, and other materials to communicate thoughts and information. Of course this worked, but the material was slow and often expensive, two characteristics that the invention of paper changed.

Papyrus, the first material that was like paper, was used by the Egyptians way back in 4000 B.C. Papyrus was made of reeds pounded together to make a hard, thin sheet suitable for writing.

Paper as we know it was invented by the Chinese in A.D. 105 by a eunuch of the Imperial Court named Cai Lin. Prior to this invention, the Chinese had written on silk, which was very expensive, or bamboo tables, which weighed too much. Cai Lin came up with a cheaper and lighter alternative. He told the court that he had created paper, a blend of bark, fish net, and bamboo that was pressed to make a very thin material that was easy to write on.

History indicates that Cai Lin had improved on another product rather than create one out of thin air. Before him, there was paper made from hemp, an Asian fibrous plant, that had been around at least since 49 B.C.

The Chinese used paper for a variety of things besides writing: for wrapping things, decorative arts, and clothing, among other things. Within a few hundred years, the new paper and thin variations thereof had supplanted silk, wooden tablets, and bamboo for writing.

Around A.D. 600, Buddhist monks spread the art of papermaking to Japan, which quickly became *the* writing material in the country as well as the raw material for dolls, fans, and dividing material for homes (i.e., screens). Around 750, the Chinese went to war with the Arabs and many of the Chinese were captured. To facilitate their freedom, they told the Arabs that they would teach them the secrets of papermaking.

It took awhile for what the Arabs learned about making paper to spread to Europe. But it did get there. The Arabs constructed the first paper mill in Xativa, Spain, around 1000 and it continued to be made by the Moors when the European armies drove them out. But this was a positive rather than negative event, because the knowledge of papermaking then spread to Christian Europe.

Until around 1250, Italy was a prime producer and exporter of paper, and then around the midpoint of the fourteenth century, French monks started making paper for use in recording holy texts. All of this was fine, but what was written on the paper was still being done with a pen, so that information, despite the availability of paper, could not be widely disseminated.

Then, the Germans started making paper (with technical assistance from the Italians) and vastly improved the craft and quality of the paper. Then in 1453 Johan Gutenberg invented the movable type printing press. Books, which were once only owned by the select few, such as royalty and the clergy, started to be accessible to everyone, including the commoners. And as people learned to read, the demand for reading material rose and, consequently, the need for paper kept apace.

Over the next two centuries, paper manufacturing spread worldwide, including to the New World. The first paper plant was founded in Mexico around 1680, and then an American named William Rittenhouse founded the first paper mill in North America in Philadelphia.

For a long time, paper was being made from old clothes, rags, and other fabrics, but slowly there seemed to develop a worldwide shortage of these materials. Then, a Frenchman named René-Antoine Ferchault de Réaumur suggested that wood could be used, after watching how wasps constructed their hives. It was a great suggestion, but turning a tree into a writing tablet was still a ways off.

This came about gradually. In 1852, an Englishman named Hugh Burgess helped make a better wood pulp, the basic raw material of paper manufacture. And two years before that, a German named Friedrich Keller devised a hand-cranked papermaking machine that turned paper out in large sheets.

The quality of the pulp was gradually improved, first in 1867 by an American named C. B. Tilghman, who added sulfites to the pulping process, and then ten years later by C. F. Dahl, a Swede who added another chemical and improved it even more. His so-called sulfate method came to America in 1907.

In 1883, Charles Stilwell invented a machine to make brown paper bags, and from 1889 to 1900 the production of paper exploded, reaching 2.5 million tons a year. Indeed, once the writing slate was used in schools, but when paper arrived this was put away for good.

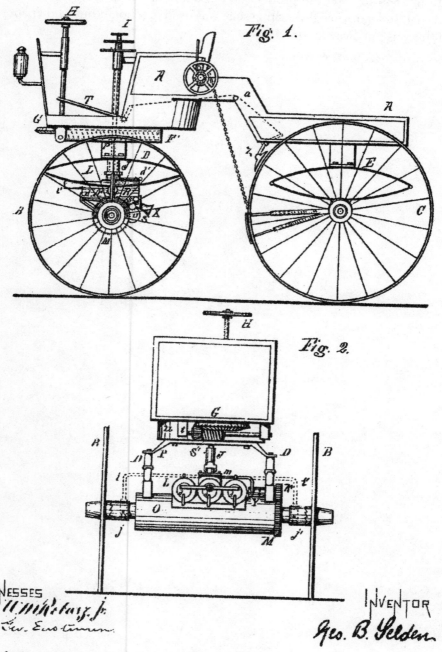

Patent drawing, 1895, by G. B. Selden. *U.S. Patent Office*

Automobile   13

Legend has it that automobile baron Henry Ford had a unique way of firing people: an employee would leave on a Friday and return on Monday to find his office cleaned out and a notice that he was no longer employed by the company.

Regardless, three things are known for certain about Ford: he revolutionized the automobile worldwide, creating a car that was practical and affordable; he brought into being something called the production line, which revolutionized industry; and he changed the way America lived. Before Ford, there was the horse and buggy. After Ford, there was the motorcar.

Members of the Ford family started to come from County Cork, Ireland, to Dearborn, Michigan, in 1832. William, Henry's father, and a host of uncles and aunts arrived in the area in the 1840s, driven from Ireland by the potato famine

Michigan was a good place to immigrate to. At the time, anyone could buy an acre of land for the munificent sum of $120. The immigrants bought every square inch of soil they could afford and then set about farming it. At harvest time, the produce was sold in Detroit, which was not so far away that it couldn't be reached by horse-drawn carts.

Henry, born two years before the Civil War ended, worked on his family's land. But when he was sixteen, he took a part-time job at a machine shop where he could exercise his interest in studying how mechanical things work and in inventing. He then went to work for the Detroit Edison Company and by the time he was thirty, he had worked his way up to being chief and was responsible for that city's electricity.

The job allowed him a lot of free time. While he had to be on call 24 hours a day, circumstances rarely required his presence. He was thus able to sequester himself in his shop where, in 1893, he constructed a gasoline engine that was an improvement over its predecessors. Three years later, he invented an ungainly spidery-looking thing with four wheels that was part bicycle, part car. He called it the "quadricycle" or "horseless carriage."

Over the next few years, he improved the horseless carriage and in 1903 he felt he had developed a marketable car. With just $28,000, Ford incorporated the Henry Ford Company

His company was a success (he publicized it by racing his car; he himself drove a "999" to a world record, covering a mile in 39.4 seconds) and was almost immediately pounced on by the Licensed Auto Manufacturers, who said he couldn't use the gasoline engine, which, they claimed, had been patented in 1895. Ford disagreed, saying his engine was different from the original. They went to court and in 1903 Ford lost. But in 1911 he won on appeal.

In 1908, Ford told the world that he would build a car for the masses, and he did. The Model T sold more than fifteen million vehicles, and Ford captured half the world market

The core of his success was not just the car, which was well made, but the value consumers received for their money In 1908, the Model T cost $950, but because of Ford's innovations on the production line and because of his willingness to pay his workers double what other auto manufacturers did, which encouraged his workers to greater productivity, in 1927 he was able to produce a Model T that sold for less than $300. To get parts to build his car, Ford bought the producers—the mines, forests, glassworks, and rubber plantations—of the raw materials needed, as well as the ships and trains to transport them. His income was now so great that he could finance these purchases himself.

Though Ford's car and production line achievement helped eliminate the old ways that things were done as well as the way people lived, he never lost a liking of things from yesteryear. To help preserve those traditions, he built Greenfield Village near Detroit, which sought to reproduce things the way they had been when he was a boy. And his admiration for Thomas Alva Edison (he once wrote in one of his notebooks, "God needed Edison") was reflected in his Greenfield Village duplication of the Menlo Park, New Jersey, laboratories where Edison had invented so much. Ford had worked for Edison and regarded him as his mentor. Early on, when he had been working on the gasoline engine, Edison had encouraged him to continue, rather than get involved with steam or some other fuel system.

In the 1930s, the fortunes of the Henry Ford Company declined. The successor to the Model T, the Model A, did not sell as well, and across the 1930s the graph line showing the company's sales continued downward. But when World War II came along and with it the demand for thousands of new vehicles, Ford rebounded.

Ford was a tough man, but the great sadness of his life, the one thing he could never put behind him, was the death of his son Edsel, who died from cancer in 1943. It is said the heart went out of him not only for business, but for life itself. Two years after Edsel's death, Ford handed the reins (or steering wheel) of the company to his grandson Henry II. He died four years after Edsel and willed his shares of the company to the Ford Foundation, instantly making it a leading philanthropic organization.

Airplane   14

Ancient tablets and drawings are replete with birdlike images—many including winged and feathered humans able to ascend and descend in the mystical skies. Since the very dawn of humanity, our greatest wish has been to soar like the birds—a symbol of freedom, grace, and mystery. Early attempts at flight relied entirely on the emulation of birds. Numerous experimenters attached feathers to their arms and legs and bravely attempted to turn the dream into a reality. Their efforts, however, usually ended in failure, as many were either killed or injured leaping from cliffs or other high places. But just under one hundred years ago, what had eluded Aristotle, Leonardo da Vinci, and Galileo was brought to reality by two bicycle mechanics from Ohio. The Wright brothers are costars in this story, but there was a supporting cast and a fascinating script.

In the nineteenth century, the theoretical groundwork was laid for powered flight. Sir George Cayley, a wealthy British philosopher, politician, and educator, pioneered the research on the elements of wing structures and the need for a light-weight power source. Cayley proposed that wing design must incorporate accommodation for drag as well as lifting capability. He also postulated that the angle at which air was moved over a wing affected lifting power.

"The whole problem," he wrote, "is confined within these limits—to make a surface support a given weight by the application of power to the resistance of air." He also made the astonishingly correct prediction that the propulsion mechanism would need to be powered by—in his words—"the sudden combustion of inflammable powders or fluids."

Cayley built a small glider, based on his research in 1804. By 1809, he successfully launched a larger, though unmanned model. He continued his studies and soon constructed another glider, incorporating a "streamlined" fuselage and a moveable tail structure. He convinced a local schoolboy to take to the air in this device for a short, successful "flight" down a hillside.

Despite his great innovations, Cayley was stymied by the technology of his times. The only available power source—the steam engine—proved unsuitable for air power application. Coming into widespread use in the 1800s, the steam engine was revolutionizing ship design and making possible the railroads. But boats and locomotives didn't need to leave the earth's surface and take to the air. Steam engines were massive and heavy—too heavy for a given horsepower ratio—and

Patent drawing, 1906, by Orville and Wilbur Wright. *U.S. Patent Office* required large amounts of either wood or coal for fuel and water for steam generation.

Cayley's glider designs, however, did not go unnoticed. Many studied and copied his efforts. Indeed, gliders are still much in use today and have logged flights of impressive distance and time. But they are, like the hot-air balloon, the slave of prevailing meteorological conditions. Attention soon turned to adopting the glider to powered flight. William Henson proposed an "aerial steam carriage" with a 150-foot wingspan and twirl propellers. In 1848, he attempted to fly a scaled-down 20-foot version that used a lightweight steam engine as power. This avant-garde prototype got off the ground, but it was hampered by the still-too-heavy steam plant, not to mention that it glided rather than flew.

In the latter part of the nineteenth century, two significant pioneers in air flight moved into the forefront of the quest for flight. Otto Lilienthal published the widely read book *Bird Flight As the Basis of Aviation,* which was based on many years of research and observations he had done of birds in flight. Lilienthal built and conducted numerous glider experiments and incorporated a small gasoline engine into his design. He was tragically killed in 1896 testing this plane. Sir Hiram Maxim built a steam-powered biplane in 1894. The plane boasted twin engines and twin propellers; it actually rose from the ground but was tethered in place with guard rails. Despite this promising start, Maxim inexplicably ended work on the project.

With the dawn of the twentieth century, the race to be "First in Flight" likewise rose over the horizon, and the new century brought the Wright brothers to the head of the list of would-be aviators. Orville and Wilbur Wright were the sons of an Ohio clergyman. The boys soon became adept at mechanics and invented a printing apparatus while still in their teens. They worked in the printing trade until 1892 when they opened a bicycle shop in Dayton, Ohio. The boys had received a present of one of Alphonse Penaud's rubber band toys in early childhood and later read Lilienthal's research papers. They believed they could improve the Lilienthal designs and correct other faults in aircraft theories then in vogue.

Their principal advance in design was "aileron control," the warping of wing surfaces by cables that would enable the plane to remain in equilibrium while making turns. The Wrights had observed how birds in flight restored balance by angling one wing down and the other up. Experimenting with a cardboard box, they could duplicate this action by warping one side and the other—changing the aerodynamic structure without sacrificing rigidity.

Many people to this day regard the Wright brothers as "hit-or-miss" experimenters who got lucky at the right time and place. Nothing could be further from the truth. They constructed one of the first wind tunnels in their Dayton shop and painstakingly tested patterns for fuselage and wing configurations. Monoplane, biplane, and even triplane models were carefully tested and evaluated. In addition, when they could not obtain a gasoline engine to meet their specifications, they built their own.

In 1900, the Wrights were prepared to put their research into action. They built a nonpowered glider incorporating their "wing-warp" design. They then proceeded to Kitty Hawk, North Carolina, a site chosen for its steady breezes and clear beach.

After a series of promising flights, they returned to Dayton where they continued further glider design through 1901 and 1902. By 1903, they were prepared to install their engine and become the "First in Flight."

The Wright brothers returned to Kitty Hawk and chose December 14 for their first attempt. Their finished plane—*The Flyer*—was a biplane with canvas-covered struts and body; it had a wingspan of 40 feet and weighed 805 pounds. Their custom-built engine was an in-line four-cylinder automobile-type plant that generated 13 horsepower and weighed 180 pounds. *The Flyer* also featured twin propellers driven by bycicle-type gears and chains.

Being the winner of a coin toss, Wilbur was given the privilege of attempting the first takeoff. The plane, however, quickly faltered after the launch and belly-landed on the beach. Although Wilbur was not hurt, *The Flyer* sustained some damage and the next attempt was rescheduled for December 17.

December 17, 1903, was to be one of the great dates in the history of the twentieth century. With Orville taking the controls, *The Flyer* took off smartly and flew under its own power for 12 seconds, covering about 120 feet. "A machine carrying a man had raised itself by its own power into the air in full flight," Orville announced to the world, "and had sailed forward without reduction of speed and had landed at a point as high as that from which it had started."

With photographers and press in attendance, they flew three more flights that day with the final one lasting nearly a minute and covering 852 feet. In 1905, Wilbur flew for over half an hour and covered 24 miles in a circular course.

America and the world honored the brothers with numerous medals and awards, but Wilbur contracted typhoid fever and died in 1912. Orville survived until 1948, long enough to see *The Flyer* change the twentieth century and revamp our concept of the world.

Plow 15

The plow is a very simple device, but it certainly deserves a high place on this list. At the core of its use for humanity is speed and efficiency in making ruts in the land so seed can be cast and crops produced. If the plow were unknown today, feeding the world's billions would be much more difficult. Indeed, in some countries it might be close to impossible.

The plow had humble beginnings, probably just a man scratching the soil with a stick to make a furrow or groove for the seed. Then, the "ard," or "scratch plow," was invented, appearing first—as archeological digs indicate—in southern Mesopotamia and dating back to 4500 B.C. It was simply a device featuring a single pole with a pointed end called a "share" that could be dragged across the ground to make a furrow. Such devices were first pulled by men, but eventually one or two oxen—who, it was discovered, were able to work all day without getting tired—were employed.

While wooden plows could work on light sandy soil such as that found in Mesopotamia and Egypt where the weather was mild and dry, they could not work nearly as well in countries where the soil was heavy and wet. Thus, the use of domesticated animals such as cows or oxen became vital.

The big development in the plow came from China, a country far removed from others and one that was quite inventive. The Chinese were also quite secretive about their world and their inventions, so the Western world did not discover what the Chinese knew by 3000 B.C., namely that sharp stones could be shaped into "plowshares" and were more efficient than wooden ones. While stone, of course, was heavier than wood, the Chinese had discovered that the shares could cut deep furrows in heavy soil.

The Chinese eventually developed an iron plowshare by 600 B.C., at least five hundred years before their Western counterparts! The iron plowshare was clearly superior. For one thing, it could be tooled into the most effective shape. For another, it was quicker to use than stone or wood.

The Chinese actually developed two types of plowshares. One was all iron, and the other was an iron share secured to wood. Because of its weight, the all-iron type was not as easy to use as the other, where the share was secured to a wooden frame.

The Chinese also developed iron to a point where it was very strong, essentially by blending molten iron with minerals that made it strong but not brittle.

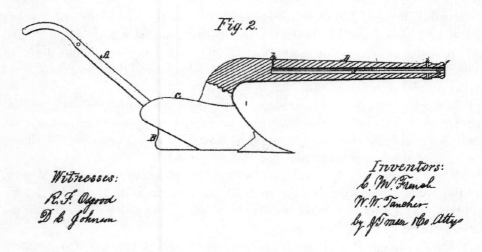

Patent drawing, 1862, by French and Foucher. *U.S. Patent Office*

Previously, when it was just made molten and poured into shape, the resulting cast iron could hit a rock and break.

Another problem was that as the plow made a furrow in the earth, the earth would fall back into the furrow, requiring the farmer to remove it. To solve this problem, the Chinese developed the moldboard, which was a curved metal plate that pushed the plowed earth well clear of the furrow.

Another improvement was to make a plow that could be adjusted to different depths, a boon on digging in different types of soil.

Europe and the rest of the world learned about the Chinese improvements on the plow in the seventeenth century, when China opened its ports to trade. Dutch traders brought information about the plow to Europe. This is indicated because the standard plow used in northern Europe in the eighteenth century was called the "Rotherham," which was the name of the place where it was manufactured in Yorkshire, England, and its origin was the Netherlands. The all-iron plow was introduced in Europe by the end of the eighteenth century. Developments that followed included plows with replaceable parts and that could be equipped with shares that were better suited to particular soils.

Until around 1850, plowing was done with the help of oxen or horses. But then steam-powered plows came into the picture. At this time, they were so expensive, however, that they were mostly owned by traveling contractors rather than individual farmers. The advantages were great. While domestic animals could pull a multishared plow on light soil, the steam-powered version could do the same thing on heavy soils and complete up to 12 acres a day.

By the end of the nineteenth century, individual, mobile, steam-powered plows were developed, and all the farmer had to do was operate the wheeled machine, pulling the plow behind him. The twentieth century saw the introduction of the internal combustion engine, and mechanized plows were equipped with this instead of the steam engine, which became obsolete.

16

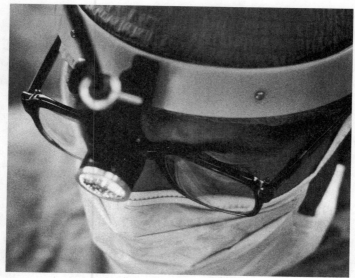

Neurosurgeon Charles Wilson. *Photofest*

Eyeglasses

It seems clear—no pun intended—that almost until the time of Christ, eyeglasses had not been invented. Indeed, in order for prominent but poorly sighted Romans of the day to read, they had to have their slaves read aloud to them.

A lens that approximated a magnifying glass and made of polished rock crystal an inch and a half in diameter was found by archaeologists near the ancient city of Ninevah, Assyria. And the playwright Aristophanes makes reference to such a glass being used with the sun's rays to burn holes into parchment and of erasing writing from wax tablets by melting the wax.

What was known as a "reading stone," which we would call a magnifying glass, was developed around A.D. 1000. Apparently, the Venetians learned how to develop such a glass, which was laid flat on the words to be read, thus magnifying them. There is reference in surviving texts to the presbyopic monks who used such glasses to read. At one point, the Venetians took the glass off the paper, as it were, and housed it in a frame that could be put in front of the eyes.

The first eyeglasses, or spectacles as we know them, appear to have been invented between 1268 and 1289. The reference to eyeglasses in 1268 comes from scientist Roger Bacon, who wrote in his encyclopedic *Opus majus* of examining "letters or minute objects through the medium of crystal or glass or other transparent substance" so that they were magnified. Then, in a 1269 paper entitled "Tarite de con uite de la famile," a man named Sandra di Popozo wrote, "I am so debilitated by age that without the glasses known as spectacles, I would no longer be able to read or write. These have recently been invented for the benefit of poor old people whose sight has become weak."

Unhappily, the name of the man who invented spectacles is not mentioned, but there is one reference to him in a sermon given by a monk in Pisa in 1306. "It is not twenty years," he said, "since the art of making spectacles, one of the most useful arts on earth, was discovered. I, myself, have seen and conversed with the man who made them first."

The first spectacles had quartz lenses for a simple reason: glass had not yet been invented.

Surprisingly, the most common problem eyeglasses had following their invention—in fact this problem plagued them for some 350 years—was how to keep them on. Eyeglasses rest on the nose and hook over the ears, but these body features vary in size and shape and in their ability to support them. Moreover, the lenses are supposed to be perpendicular to the visual axis, but this is only possible when the eyes are trained in one direction.

A variety of frames were created to keep the eyeglasses on, and in 1730 a London optician named Edward Scarlett perfected the use of rigid sidepieces that hooked over the ears.

There were other developments, including the use of colored lenses because some inventors felt that plain glass allowed for too much harsh light to pass through them. Hence, some lenses were yellow, green, turquoise, or blue.

Different cultures had different attitudes toward using eyeglasses. For example, the French and English used eyeglasses in private, while in Spain the attitude was completely different: Spanish society felt that eyeglasses made one look more important and dignified.

In America, cost was a prime consideration of who used eyeglasses. While they were for everybody, the whopping $200 price tag, which would be equivalent to thousands of dollars today, put them out of reach for many people.

While Benjamin Franklin is remembered for many things, his role in the development of eyeglasses is not that well known. He was the inventor of bifocal eyeglasses, where one section of the glass is used for seeing things close by and another for seeing things far away. Franklin had invented them because he found using two separate pairs of spectacles cumbersome. As he said,

> *I . . . formerly had two pairs of spectacles, which I shifted occasionally, as in traveling I sometimes read, and often wanted to regard the prospectors. Finding this change troublesome, and not always sufficiently ready, I had the glasses cut in half and a half of each kind asso-*

ciated in the same circle. By this means, as I wear my own spectacles constantly, I have only to move my eyes up or down, as I want to see distinctly far or near, the proper glasses being always ready.

There were some problems with bifocals, including dirt gathering at the point where the the two lenses were cemented together, but eventually it became possible to make them out of a single piece of glass.

Contact lenses have a surprisingly long history, the idea first emerging in 1845 when a man named John Hershel suggested them. Like so many other inventions, they were born of necessity. As it happened, late in the nineteenth century a man whose eyelids had been destroyed by cancer had contact lenses invented by F. E. Muller, a German who made glass eyes, placed over his eyes. The lenses lasted until the man died twenty years later.

The first contact lenses were large and relatively uncomfortable, but as time went by and materials improved, they became thinner, smaller, more comfortable, and, of course, more popular. By 1964, over six million people were wearing them, 65 percent of them women.

Humanity's ability to see better is of incalculable worth. Just imagine, if you will, what other inventions and advances in science would not have happened without them. Eyeglasses were a simple invention. So, indeed, was the wheel.

Atomic Reactor 17

Atomic power is, of course, the greatest power that human beings have thus far been able to harness, and the men—Enrico Fermi and his partner Leo Szilard—who enabled this accomplishment did so by inventing a nuclear or atomic reactor.

Their achievement was to control the energy released by neutron-bombarded uranium in a chain reaction. They were granted the patent in 1955, but it was assigned to the U.S. government because Fermi and Szilard had worked for it during World War II developing the atomic bomb.

Fermi was born in Rome, Italy, on September 29, 1901. He had always been interested in mathematics and physics, and as he got older he had an engineer as a mentor. He became so knowledgeable in these subjects that he was awarded a scholarship in 1918 to attend the Scuola Normale Superiore of the University of Pisa. He graduated magna cum laude four years later with a doctorate in physics. He became a professor of physics in Rome, then atomic physics, in particular the creation of artificial isotopes through neutron bombardment. His work in the field was so significant that in 1938 he was awarded the Nobel Prize in physics.

It was during this time that Fermi suffered for his antifascist views as well as because his wife happened to be Jewish. Fermi seized the moment. When he went to Stockholm, accompanied by his wife, to receive his Nobel Prize, he never returned, having obtained a position as a physics instructor at Columbia University in New York.

It was at Columbia that Fermi teamed up with Szilard and a graduate student and researcher named Walter Zinn to conduct experiments on nuclear fission. The group concluded that enough neutrons were released during the process to cause a chain reaction—that is, a release of energy.

It was seen to be of potential military significance, and in March 1939 it was arranged for Fermi to talk to the U.S. Navy about the phenomenon. Even though the navy seemed intrigued by what Fermi had to say, nothing came of the discussion.

A few months later, Szilard explained to Albert Einstein the kind of work the investigative team was doing, and Einstein, who carried a lot of weight in political circles, told President Franklin D. Roosevelt. Roosevelt followed through, getting other scientists involved in the work, which was deemed to be quite significant. Near the end of 1940, Columbia was given a grant of $40,000 to explore the sub-

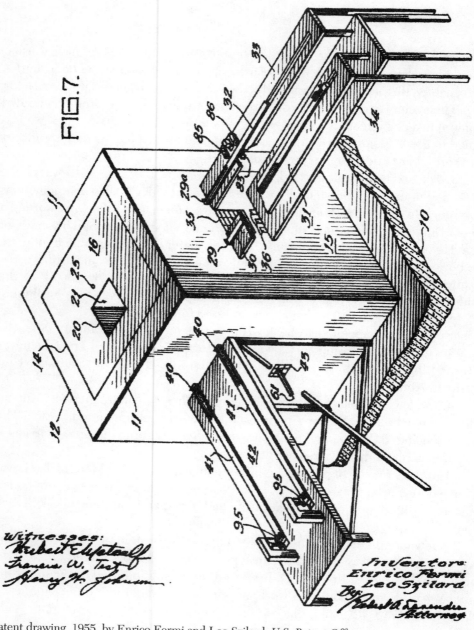

Patent drawing, 1955, by Enrico Fermi and Leo Szilard. *U.S. Patent Office*

ject even more deeply and to set up conditions required for controlled nuclear fission by a team that was led by Fermi.

The work of Fermi's team, in turn, was monitored by a team of scientists from Princeton University. They approved of it, and a year after hostilities erupted between the United States and the Axis Powers the team set up a sort of super investigative team in Chicago.

On December 2, 1942, the group was ready for a momentous experiment. It was conducted on the floor of a squash court under the stands at Stagg Field at the University of Chicago and was the first attempt to control a chain reaction of nuclear fission.

The experiment was a success.

Work continued apace over the next couple of years. Then, in August 1944 the operation was moved to Los Alamos, New Mexico, where a new laboratory had been constructed and was headed by J. Robert Oppenheimer. Fermi became chief of the physics department. There was never any doubt as to what the goal of the laboratory was: to build an atomic bomb.

It took the team, with the war still raging, another year—and $2 billion—to try to set off a bomb. This they did at 5:30 A.M. on July 16, 1945, in an isolated section of Alamogordo Air Base, which was 120 miles southeast of Albuquerque, New Mexico, the event witnessed only by military personnel and scientists.

It was a success, and the team wasted no time in preparing atomic bombs for use in the war. This occurred only three weeks later. On August 6, 1945, the first bomb destroyed Nagasaki and, a few days later, the second was dropped on Hiroshima.

18

Bikini Atoll, July 24, 1946. *Photofest*

Atomic Bomb

The invention of the atomic bomb was a turning point for humanity. After dropping two of them on the Japanese cities of Nagasaki and Hiroshima in August 1945, America abruptly put an end to World War II and began an unprecedented time of prosperity and peace.

Some believe the price humanity is paying for such an invention is steep. It includes a terrifying scenario. We now know the atomic bomb could be a tool of humanity's destruction, and this knowledge has influenced the policies of lawmakers, government bodies, and individual psyches throughout the Cold War to the present.

In the beginning, America seemed determined to stay isolated from the war that had started in Europe while Germany was devouring various countries. But then there was the Japanese "surprise attack" on Pearl Harbor on December 7, 1941, and before the United States could mobilize, most of Southeast Asia had fallen to Japan. But, as one Japanese general was heard to comment, "I am afraid we've awakened a sleeping tiger." Slowly, America took back Japanese-occupied territory in the Pacific and stopped Japan. But America believed that the real and growing threat was Germany.

But this was all-out war, a fight to the finish, and countries, including the

United States, would use whatever they could to win. Under the guidance of President Franklin D. Roosevelt, a top-secret joint effort between America and the United Kingdom was begun to build an atomic bomb. Headed by General Leslie R. Groves at isolated locations such as Los Alamos, New Mexico, the program was known only to a small number of scientists and politicians. Indeed, President Harry Truman only learned of the project, dubbed "The Manhattan Project," from Secretary of War Henry Stimson on April 25, 1945, only after becoming president, not during his tenure as vice presdient.

America knew it faced fanatical foes. Indeed, in the event of an American invasion, Japan was said to have trained hundreds of children, with dynamite strapped to their bodies, to throw themselves under American tanks. Suicide bombers had been used throughout the war and would continue to be used. It was a believable scenario.

The Manhattan Project was by no means sure to succeed. As it happened, there were numerous managerial and technical difficulties, and the experts knew that the creation of the bomb rested on a theory that was completely unproven.

Indeed, by the beginning of 1945, $2 billion had been spent on the project, and up to a very late date there were doubts as to the best method of detonation. Also, while the bomb was being built, there was considerable pressure not only to maintain safety, but also to keep the rest of the world from learning of the bomb.

The key element in the making of the bomb was the creation of plutonium, which does not exist in nature, but like uranium (which does exist in nature), when given the right stimulus, could produce a chain reaction. This runaway reaction, which is what makes the atomic bomb "work," is called "fission" and was investigated and made comprehensible to others by a physicist named Niels Bohr.

Fission occurs when the nucleus (the central part of an atom) breaks up into two equal parts. The nature of atoms is predictable and dependable for this process, and once a neutron breaks up the uranium atom, the fragments release other neutrons that break up more atoms, and so on.

This chain reaction takes place in only millionths of a second and the amount of power released during this reaction is several hundred million volts of energy. During fission large amounts of heat and radiation are given off—and the radiation produced is called gamma radiation, the deadliest form known to humans.

The first atom bomb was exploded in the desert of New Mexico and was code named *Trinity*. The bomb was placed inside a huge steel vessel named *Jumbo*, which was 6 meters long and weighed 200 tons. The bomb that was to be dropped on Nagasaki was called *Fat Man*, but still comparatively tiny. It consisted of a hollow sphere of plutonium encapsulated in layers of fast- and slow-acting explosives.

Detonators would ignite the explosives, which produced a symmetrical shock wave that collapsed the sphere, "setting off" the plutonium and resulting in the nuclear explosion. The key to the detonation was the explosives, which burned at different rates. In this way, the shock waves would focus on the sphere in much the same way as a lens focuses light rays.

The bomb that was dropped on Hiroshima contained two subcritical masses of uranium (a nuclear reaction occurs when the uranium goes to "critical mass"). At ground zero, one mass of uranium was fired into the other to produce the explosion.

Years later, many American soldiers who took part in the bombings were said to have second thoughts about the havoc that these bombs wrought, but looking back there were others who would say that they would do it again if called on, because the Japanese gave no one any other choice.

Colossus Computer 19

Many people have no idea who Alan Mathison Turing was, yet his influence on our lives has been nothing short of profound.

Turing was a mathematical and scientific wizard, a pioneer in computer science. Military historians say that his work in breaking the so-called Enigma code machine (which was a kind of computer) that the Germans used during World War II shortened the war by two to three years, and thus saved untold numbers of lives—perhaps millions—and averted other havoc. Historian Sir Harry Hinsley, who worked during the war interpreting naval messages that were sent to him by Turing, speculates that if Enigma had not been broken, the invasion of Europe might not have taken place until 1946 or 1947 instead of June 6, 1944. Also, after July 1945 American atomic bombs were being produced at the rate of one a month and were available for wiping out German cities or submarine bases—and contaminating those areas. A protracted war might also have resulted in a guerrilla war by driving the German army into the hills. Germany by then was also perfecting the V2, an early type of missile that was directed at Britain, and would eventually have had chemical and biological weapons at its disposal. If the invasion took place in 1946 or 1947, would the Soviet union have then allowed Germany to surrender to America? If it didn't, what would the political implications have been?

All those things never occurred, in large part because of Turing. He might have been feted, indeed adored and revered, perhaps a statue erected for him in downtown London. Instead, on June 9, 1954, at the ripe old age of forty-two, Turing, his heart broken and his mind in disarray due to the loss of his security clearance—because he was a homosexual—sprinkled potassium cyanide on an apple and ate it, ending his life. And with him died, of course, his transcendent intellect and whatever else he might have given to humanity.

Turing was born in London on June 23, 1912, and his scientific genius and interest in things scientific emerged early, but he showed little interest in history, Latin, English, and the like. In 1931, he entered King's College, Cambridge University, where he focused on mathematics and developed an abiding interest in re-creating the work of other scientists.

At one point, he started to develop a kind of digital computer, which was dubbed the "Turing machine." The machine read a series of ones and zeros from a tape, which described the particular task that had to be done. The key was to in-

...ict the computer properly and then it would perform the task or tasks. He believed that an "algorithm" could be developed to solve any problem. The only difficult part would be breaking down the problem into steps the computer could recognize.

In the 1950s, there were computers, but most were designed to do a single task. Turing's concept was regarded at the time as outlandish, rather than revolutionary. Today of course, what he described is exactly what programmers do.

During the 1920s, the Germans created the Enigma code machine, which led them to believe that their coded messages concerning military and other top-secret operations were beyond being decoded. It was a reasonable assumption. The machine, which resembled a typewriter, was capable of doing millions of calculations in milliseconds, and the secret codes that controlled them were changed at the start of every day.

In the early 1930s, however, Polish mathematicians had obtained a machine and started to try to break it, because they felt that one day they would be invaded by Germany and it could be a great asset. Turing took it the rest of the way.

He headed up a team of scientists and mathematical wizards assembled in Bletchley Park outside London whose specific goal was to try to break Enigma's code. To do so, the team developed a computer—perhaps the first computer—called the "Colossus." It operated with 1,500 vacuum tubes and worked 24 hours a day. As time went by, improved models were installed and though it is still a secret today, experts believe that a total of ten Collossuses were built. Regardless, the fact that it helped break the Enigma code was the breakthrough that instantly became the greatest secret of World War II. In practical terms, it meant that the Allies knew exactly what the Germans planned to do before they did it, an incalculable military advantage. How important? It helped the Allies decide where to invade on D day—and fool Adolf Hitler in the process.

Nothing mattered as much. At one point, Bletchley Park alerted Prime Minister Winston Churchill of the imminent bombing of the city of Coventry, and he let it occur, with much destruction and loss of life, rather than clear the city and risk having the Germans deduce how the British had known about it. (Throughout the war, the British used an elaborate system of counterfeit spies and surveillance to keep the Germans from figuring it out.)

At another point, though, a major crisis arose. The Germans started using a variation of Enigma to direct their submarines, one that amounted to a brand new code. What Turing and his colleagues knew had abruptly become useless; now they could not warn anyone of anything.

The result was terrifying. Scores of people as well as tons of Allied shipping went to the bottom of the ocean and Turing and colleagues could do nothing about it except try, feverishly, to break the new code.

And, finally, in as dramatic a scenario as one can imagine, it came down to Turing himself, desperate, alone in one of the cottages at Bletchley and working to exhaustion. His magnificent mind eventually broke it, and the shipping was subtly rerouted out of harm's way.

After the war, Turning worked on a variety of machines that would supplant human intelligence. Indeed, he believed that a machine could be developed that replicated human intelligence, the inspiration said to be the loss of a young love in his life. Turing was literally trying to bring his love back to life. He also wrote a paper in 1950 containing what is today known as the "Turing test," which evaluates a machine's intelligence, a test still considered the standard by which mechanical intelligence is evaluated.

While his homosexuality was not an issue during the war, after 1948 homosexuality in general became more looked down on as the political and emotional landscape changed with the development of the Cold War and Britain's alliance with the United States. Regardless, the machine he helped build stands as one of the greatest inventions of all time.

20

Modern American toilet. *American Standard*

Toilet

The toilet is a crucial device in the history of improving human health. Although most cultures consider it taboo to talk of bodily functions, throughout much of human history the lack of excremental hygiene was a national health hazard. Put in other terms, sanitary practices for the removal of human waste has always been needed. The toilet was the answer.

Still, like many other inventions, the modern toilet's history was riddled with setbacks and delays. Different cultures made strides during some time periods and fell back in other periods.

For example, there were toilets connected to drains made of clay bricks in India as far back as 2500 B.C. Still, during the Dark Ages, from A.D. 500 to 1500, people were emptying buckets and chamber pots filled with human excrement from their windows. Open cesspools were common, disease was rampant, and people literally died in the streets. The situation in Europe followed a similar pattern for an astonishingly long time throughout human history.

Before having toilets inside houses, people were forced to remove human waste in other ways: burying it in the woods, throwing it out windows into open sewers, dropping it into moving rivers, or using chamber pots that had to be cleaned periodically.

The public's habits and attitudes were also important in shaping the future of toilets. Although there were pockets of people throughout history that experimented with toilets on a small scale (usually the rich), it wasn't until the sixteenth century that governments began to act on the need for sanitary conditions. The assumption was that dirt equaled disorder, which was bad for society. Still, people continued to use the outdoors to dump waste. Although laws were passed from the sixteenth century on to make toilets in each house mandatory and public toilets available, it wasn't until the eighteenth century that real, steady progress was made.

Although John Harrington invented the water closet in 1596 (much like modern toilets but the water supply was contained in a tank that resembled a closet and located above the toilet), it wasn't until more than 180 years later that it was adopted on a large scale. At this time, the toilet was entering private homes, but it remained crude by today's standards.

The world saw the development of earth closets and pan closets. The earth closet was a hole that was covered over after the user relieved him- or herself. The pan closet had a deeper hole that had a lid that closed the excrement off from the outside. These were effective, but they still required manual cleaning.

A turning point came in 1738 when J. F. Brandel introduced the valve-type flush toilet. After this, Alexander Cummings improved the technology and came up with a better toilet in 1775. His toilet retained water in the bowl when not in use, thereby suppressing odors, and could carry waste out of the house.

Still, the mechanics of the valve and a reliable inlet of water (the same amount and speed it entered) needed improvement. In 1777, Joseph Preser provided the required improvement and later, in 1778, Joseph Bramah substituted the slide valve with the crank valve. The technology of the pour flush (using water and gravity to "flush" remains out), had reached its peak. Finally, in 1870 S. S. Helior invented the flush-type toilet, called "optims."

From 1890 to the present, the only improvements have been aesthetic. The overall design and how toilets function remain the same. In France and England, they were placed inside the house; private bathrooms with stalls or curtains for privacy came into vogue. The other changes focussed on the shape and design of toilets (to suit personal tastes) and the amount of water being used. Conservation of water became important and designs were developed that functioned properly but with less water.

Corresponding to this was the invention of toilet paper. Prior to this, people used hemp, newspaper, and other things. Finally, in 1857 Joseph Cayetty invented toilet paper in the United States. The invention allowed people to have a tissue paper that was convenient to use, was absorbent, and within reach when needed. Although a seemingly simple device, the toilet took some time to arrive and when it was invented, it worked reliably quickly.

21

From *100 Rifles* (1969). *Photofest*

Rifle

The development and perfection of the rifle, a long-range, portable firearm, took centuries. While most of us think of it as a long gun, the term "rifle" specifically applies to the barrel of a gun—any gun or cannon really—that is bored or drilled in a spiral pattern on the inside. At first, the bore of a gun or a cannon was smooth, allowing a projectile to slide in easily and quickly for loading. Through the study of ballistics, however, it was found that a spiral or "rifled" bore could make the projectile spin through the barrel, which increased its range and accuracy.

The "hand cannon" as it was known, appeared on battlefields in the middle of the 1400s. It was a great monster of a device that weighed up to 25 pounds and required a forked brace to prop it up for the gunner to fire it.

A projectile, which was roughly spherical in shape, was rammed down the barrel on top of a measure of powder. At the bottom of the barrel was a "touchhole," an opening that led to a "flash pan" or small, saucer-shaped pan that held another measure of powder. The powder was ignited by hand to fire the shell.

The first mechanism devised to automatically bring ignition to the flash pan was called a "serpentine," a German invention that used an s-shaped piece of metal with a pivot in the middle. When one side of the serpentine, or the "trigger," was pulled by the operator, the other side, which was equipped with an adjustable set of jaws that held a slow-burning fuse called a "match," was brought to bear on the flash pan.

The next important advance was the elongation of the stock and providing it with grips for the hands. The butt itself was enlarged, so there would be less recoil that would impact against the user.

The collective result was known as the "harquebus," the precursor to the "musket," which remained in use a long time. Soldiers originally placed such a gun to their breastbone for firing, but as gunpowder got stronger, this became painful, so they switched to the shoulder, which absorbed the recoil better.

Innovations in weapon design began to slowly change how warfare was conducted since the projectile could penetrate even the best of armor. The days of the knight in shining armor began to dwindle.

The "wheel lock," a firing mechanism, was another advance. This was a metal wheel that was rotated by a spring against a spring-mounted piece of iron pyrite. When the trigger was pulled, the wheel turned and rained sparks from the pyrite into the flash pan.

By the mid-1500s, the rifle used a "flintlock," which incorporated a much simpler, cheaper design. This had a "cock," a metal jaw holding a piece of flint, that slammed into a piece of steel when the trigger was pulled.

It was the Industrial Revolution, with profoundly improved production methods, that made the rifle the battlefield scourge it was to become. The simplicity of the flintlock's design made it possible to standardize patterns, making not only mass production possible, but also repairs with interchangeable parts. As the armies of Europe grew, the industrialization of war made it possible to arm each soldier with the new weapon. And the allure of the rifle was obvious: it was a powerful, accurate, and long-range weapon that could outdistance enemy fire.

As time went by, improvements were made. First was the "percussion" firing method that brought a "hammer" or "striker" down onto a cap of fulminate of mercury, which exploded on impact. This replaced the "flint" in the flintlock. Ironically, while many experimented with the idea of the percussion cap, Alexander John Forsyth, a Scottish minister, was the first to make it a reality and had his system patented in 1807. Handily, the caps were easily adapted to preexisting standardized flintlocks.

The next important step was making the bullet small enough to be loaded easily into a rifled muzzle and then, once in the breech, expand to fit into the spiraled grooves for an optimum flight toward its target. A number of methods were tried, including musket balls that were pressed into the breech and pounded with the ramrod to make them expand at the bottom to fit the rifling. While they were much better, accuracy was forfeited because they were badly misshapen during the loading process.

Claude-Étienne Minié, a French captain who was inspired by experiments with a longer, cylindrical projectile, worked with the design to not only get around the basic loading problem, but also improve the overall performance of the projectile. His idea consisted of a hollow-based "bullet" with an iron plug in it. As the weapon was fired, the iron plug would expand and engage the rifling tightly. The cylinder of the bullet, with a dense metal cone on the tip of it, would begin to spin its way out of the barrel, having an effect like a top or gyroscope that would aerodynamically guide it to the target with unprecedented force and accuracy. As soon as Minié's invention made an appearance on the world stage, many people began to capitalize on it.

In 1851, the British sent a pattern of it to their smiths at the Royal Arms Factory at Enfield. It was in action by 1854, during the Crimean War where a *Times of London* correspondent described it as the "King of Weapons." The Russian army, equipped with the traditional smoothbore muskets, never stood a chance against the Minié as it tore apart their neat ranks and formations "like the hand of the Destroying Angel."

The Minié and the "breech-loading" system came into being just as the U.S. Civil War erupted. On September 17, 1862, at Antietam Creek in Sharpsburg, Maryland, Union troops repelled an invasion by the Confederate army. An American version of the Minié was used by the Union to face the Confederacy's own version supplemented by imported Enfields from Britain. That single day recorded over twenty-six thousand casualties on both sides and remains to this day the bloodiest in American history at the hand of a "Destroying Angel."

While the rifle continued to advance in terms of automatic loading mechanisms that allowed the soldier to fire off several rounds of ammunition in rapid succession with ever improving accuracy, it was some of these very innovations that led to the device that usurped the title of "King of Weapons." As armies continued into the twentieth century to advance on their enemies, now safely entrenched and protected from their bullets, the "machine gun" appeared on the world stage and rendered the most powerful of rifles useless.

Pistol   22

Originally, the pistol came about as a cavalry weapon, a one-handed firearm that left the other hand free to hold the reins of the horse The pistols of the era between the 1400s and 1700s reflected mostly the technology available to muskets—a single shot, muzzle-loading affair—and were initially developed as a supplement to more powerful shoulder arms.

Innovations in firing mechanisms like the "wheel lock" and "flintlock" and later "percussion ignition" systems were essential to carrying a pistol at the ready into battle. Prior to this, in order to ignite the powder charge, the user needed to have a constantly burning cord called a "match" to fire a gun. The wheel and flintlock mechanisms created a spark in the flash pan of the firearm, thereby allowing the user to keep it holstered until needed. The lightness and close-quarters effectiveness were readily apparent in the pistol's use and it became an indispensable weapon of personal defense.

Of course, as was the case with the rifle, the dream of the soldier in the field was to be able to fire off more than one shot in rapid succession. His life could depend on it.

The concept for repeated firing for both the rifle and the pistol was in place and many gunsmiths put themselves to the task of making it a reliable reality. Weapons with several barrels were created, but they lacked portability. Multiple chambers were also tried, but misfires—that is, the setting off of adjacent rounds—became a potential hazard.

As firing mechanisms improved, a gun with a revolving cylinder that brought several chambers in line with the barrel, one at a time, and with relative safety emerged. The "revolver" often possessed five or six chambers into which a ball and powder were loaded from the front of the cylinder. The soldier lined up each individual chamber with the barrel and then placed a percussion cap over a nipple that directed the flame of ignition into the chamber, ignited the powder charge, and propelled the ball out the muzzle. The percussion cap was set off by the soldier when he cocked a hammer that smashed down on the cap when the trigger was pulled.

Samuel Colt, the man who perfected this system and whose name for it became a household word, conceived of his design as a young sailor. In 1835, in Britain, France, and then the United States, he patented his design for the "per-

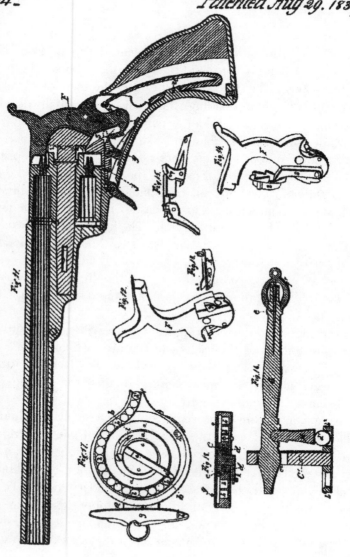

Patent drawing, 1839, by Samuel Colt. *U.S. Patent Office*

cussion revolver," which came to be called the "cap and ball" or simply the "Colt revolver."

There were two important factors that set the Colt revolver apart from other revolvers. First, Colt's mechanism allowed the revolving cylinder to move the next chamber into place as the hammer was pulled back. Not only did this create a reliable mechanical means of lining up the chamber with the barrel, but it also eliminated a step to firing a pistol. Previously, cocking the hammer and moving the chambers were done in two separate moves.

Colt's second stroke of genius in his design was in its manufacture. With the help of none other than Eli Whitney Jr., the son of the inventor of the "cotton gin" and the great American manufacturing pioneer, Colt's Hartford, Connecticut, factory was able to produce the revolver with precisely machined, fully interchangeable parts, using an assembly line manned by workers instead of craftsmen.

While the military advantages of the revolver were fairly obvious, American society at the time created another market for the rapid-fire handgun that Europeans lacked: the Wild West. As soon as word got around about how successful the revolver was in skirmishes with Native Americans in both Florida and Texas, the demand for it was mind boggling.

But in 1857, Colt's patent expired and opened the field for competition. Colt relinquished his role as king of the handgun to Americans Horace Smith and Daniel B. Wesson as they manufactured a design they purchased from Rollin White.

White's revolver used rim-fired copper cartridges all in one firing unit. This made it possible for the gun to be loaded from the rear of the cylinder instead of the front and eliminated the percussion cap, significantly cutting reloading time. Smith and Wesson improved White's design by having the revolver discharge spent casings as well as linking the action of the trigger to the hammer and cylinder. When the trigger was released, the hammer cocked and the cylinder rotated.

The automatic pistol was born in the 1890s even as its cousin the rifle was learning to load itself automatically. A few different types of designs were employed, including the "toggle link" and the "slide." Both kinds, being that of the German Luger and the American Browning, respectively, amazingly used the force and action of the recoil to eject the spent casing and to move a fresh bullet up from a spring-loaded clip housed in the hand grip. These pistols came to replace the revolver as the military's choice sidearm, which was still used as a vital, close-quarters defense weapon.

While the pistol continues to be the ultimate personal defense, guaranteed by the U.S. Constitution, since the late 1960s gun ownership in the United States has increased exponentially, resulting in an unbelievable rise in gun deaths. Debate and legislation over the role guns play in our society is likely to go on for some time. In the mean time, the Brady Bill has made some strides in controlling the proliferation of handguns. But even as the Brady Bill was enacted in 1993, almost forty thousand gun-related deaths were tallied that very same year, of which almost six thousand were people under the age of nineteen.

23

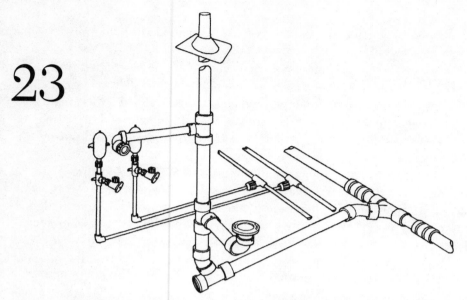

Without plumbing, adequate sanitation would be impossible. *Genova Plumbing*

Plumbing

To gauge how important plumbing has been to our lives, one can just imagine what life would be like without it.

But there is much more to it than simple convenience. Without plumbing, it is hard to envision how skyscrapers and other tall structures could have been constructed. And without plumbing, which is technically known as sanitary plumbing, there would be a greater chance of disease spreading.

Plumbing has been around a long time. Twentieth-century archaeological digs where ancient Crete once was located—it was buried under earthquakes and possibly volcanic eruptions—have produced items that indicate that the principles of hydraulics and the concepts of fundamental plumbing were well known.

Archaeologists, for example, unearthed rudimentary but highly efficient water pipes in the palace at Knossos. These consisted of hollow pieces of terracotta shaped like telescopes. The pieces were joined by sticking the narrow ends of the pipe into the wider ones, with the joints sealed with a clay cement. The pipe could be made as long as needed; additional strength could be achieved by lashing the pieces together with rope wrapped around knob-like projections. The water ran by force of gravity, and as it flowed, its action created a turbulent stream that washed away any sediment that might otherwise accumulate.

The same dig produced evidence of what is known as a drain-waste-venting (DWV) system, designed for the sanitary elimination of waste, with such features as traps, which are used to trap pockets of water so sewer gases and vermin can't get into the system, and vent pipes, which allow harmful and explosive gases to escape into the atmosphere.

But it was the Romans who brought plumbing to a true level of sophistication. Indeed, the word "plumbing" was coined during Roman times. It comes from the Latin *plumbum*, meaning "lead." Originally, and for hundreds of years, pipe was made of lead, so it was logical to call the person who worked with lead a "leader" or "plumber."

The Romans developed both a water-supply system and a DWV system throughout their empire. In the fourth century, Rome boasted almost 900 public and private baths, 1,300 public fountains and cisterns, and 150 toilets, all flushable.

Rome was a thirsty city, and to bring in nearly 50 million gallons of water to meet the city's daily needs, a 359-mile-long network of aqueducts was created, some of which, unbelievably, modern Rome still uses. The aqueducts operated both above and below ground, the water borne along in conduits until it was fed into smaller pipes (usually made of lead) that were buried underground.

As time went by, the hazards of ingesting lead were discovered and lead plumbing was banned. Lead solder was also used in sealing joints in copper pipe and was only banned in the United States in 1958. (Not all water is acidic enough to leach lead out of it, but consulting with the local water company is a good way to find out what precautions, if any, are necessary in dealing with water supply pipes that might have lead joints.)

The Roman plumbing system fell on hard times when Rome was overrun by barbarian hordes who had little interest in plumbing and no one skilled enough to maintain it when it broke down. From Roman times on, plumbing itself went into decline, borne along by beliefs that bathing, for one reason or another, was wrong. In the Dark Ages, for example, abstaining from bathing came to be regarded as a suitable mortification of the flesh of the unholy person: the use of hot water, specifically, was condemned as self-indulgent. (Some noblemen and medieval kings did bathe occasionally. In the thirteenth century, for example, it is reported that King John of England bathed at least three times a year. Queen Elizabeth is reported to have taken a bath at least once a month "whether she needed it," one of her ministers said.) The most important antibathing concept during those days was simple: bathing was dangerous to one's health.

Even in colonial times, bathing was not all that popular. Indeed, in a few fledgling states—Ohio, Virginia, and Pennsylvania—legislation banned or restricted bathing. Benjamin Franklin was an exception to the rule, as he bathed regularly. His peers, still wedded to European nonbathing traditions, criticized him, giving him that dirtiest label, "Father of American Bathing." Bathing only became popular in America in the nineteenth century.

While the concepts used in plumbing have remained the same for many years, the materials have changed. Once not only lead but galvanized steel pipe was popular, but this has been discarded and today copper and plastic pipes are very popular, as they are used for both waste and water pipe. When the man told Dustin Hoffman in *The Graduate* that the future could be boiled down to one word—"plastics"—he could not have been more accurate when it came to plumbing.

Iron-to-Steel Process   24

Steel is one the most important materials ever made, because without it so much of the world, particularly big cities, would not exist. Indeed, skyscrapers, railroads, bridges, and other important components of the infrastructure of cities would not be possible because iron is not nearly as hard or durable, so it would not allow the same structures to be erected.

Steel is manufactured from iron, and the story of its manufacture is really the story of how to control the carbon content of iron, which directly affects the material's strength and durability. If the iron contains 0.3 to 1.7 percent carbon, it is considered steel. If the iron has a carbon content below 0.3 percent it is known as "wrought," or "malleable iron," and is too soft and bendable for many of the manufacturing jobs steel can do. If the carbon content is above 1.7 percent then it is cast iron, which is heavy and strong and is commonly used to make bath fixtures—but is also brittle. A direct hit can cause it to collapse or break, hardly a plus if, for example, it was used to form the framing of a skyscraper!

Iron ore itself was first manufactured by placing the ore in a bed of heated coke or charcoal, which resulted in the carbon in the ore reacting with oxygen to create a gas that wafted away, leaving the ore behind. Early iron makers had no idea that a stronger, more durable material could be made from iron ore, but as the centuries went by, men gradually discovered that it was possible and small amounts were made.

During the eighteenth century, there was significant advancement in the manufacture of steel. In 1750, a Swedish metallurgist named T. O. Bergman discovered the importance of carbon in steel, increasing the knowledge of inventors greatly.

But it was still not mass-produced. That had to wait until two inventors, an Englishman named Henry Bessemer and an American named William Kelly, working independently, took another approach

Kelly was born in 1811 in Pittsburgh. As he grew up, he became interested in metallurgy (the science of metal) because Pittsburgh was becoming important in manufacturing iron. Indeed, when Kelly was young 10 percent of the city's population was employed in its manufacture.

Kelly got waylaid from his involvement with iron for a while, getting into the

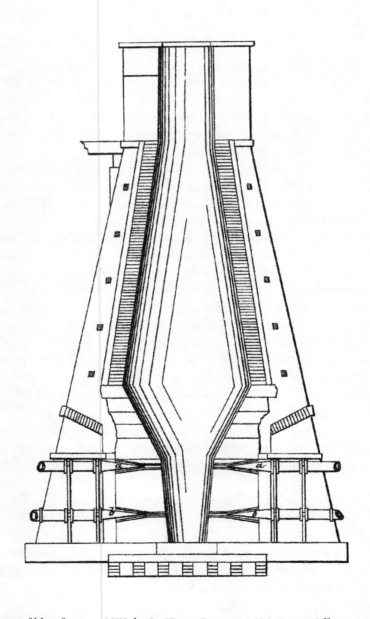

Patent drawing of blast furnace, 1856, by Sir Henry Bessemer. *U.S. Patent Office*

dry-goods business with his brother, but his travels brought him to Eddysville, Kentucky, which he couldn't help but notice was rich in iron ore. Subsequently, he and his brother formed, in 1846, the Suwanee Iron Works and Union Forge to make sugar kettles.

Kelly's business was a success, and one of the results was that he learned a great deal about the process of making iron. He began experimenting with ways to manufacture iron while using less and less charcoal, which was necessary but relatively scarce. He finally settled on a process that involved blowing a blast of cold air into molten pipe iron, allowing wrought iron to be made without any charcoal at all. Also, Kelly discovered that if he cut off the cold air at a certain point the amount of carbon in the iron would turn it into steel. He dubbed this discovery the "pneumatic process."

Kelly patented his process out of pure fear in 1856, having heard Henry Bessemer—who was to be knighted for his work—had patented the same invention. Bessemer had discovered it when he invented a rotating shell for use in cast iron cannons during the Crimean War. As it happened, the cast iron cannons were not strong enough to allow the shells to pass through without the cannon bursting, so he had naturally turned his attention to producing a stronger type of cast iron. To do this, he invented a "converter" and was able to produce a mild steel that was superior to the cast iron.

Eventually, the two men were in conflict over who owned the rights to the process, but the larger question was how to use their processes to produce a lot of steel quickly. Neither process allowed that.

Salvation came in the form of Robert F. Mushet, a Welsh metallurgist. He learned that if small amounts of spiegeleisen (an alloy of iron), carbon, and manganese were added to the wrought iron, which resulted from the Kelly-Bessemer processes, the carbon content would be elevated to the point required for steel.

Ultimately, Kelly combined with Bessemer and Mushet to create the needed technology to produce steel quickly and in volume. There were other lawsuits from other people who had created steel making companies, but by 1866 all differences had been worked out and the Pneumatic Steel Association was formed, and it wasn't long before America had surpassed England in the production of steel.

25

Wire comes in many forms. *Author*

Wire

Since antiquity, the rigging of ships and many other man-made objects were made, moved, and lifted with rope, which was made from plant (roots, vines, or strips of bark) or animal (sinew, fur, or skin) material. Wire was later developed as a stronger substitute, but it was just one of this material's uses. Wire was developed to answer a need: the ability to support and/or drag heavy loads. Later, wire became invaluable for the transmission of electricity and sound.

In the beginning of wire manufacturing, it was originally made by forging and hammering strips of metal into long strands. The method known as "drawing" was developed around A.D. 1000. The process involved pulling, or drawing, metal out to form thin, continuous strips, which produced a stronger wire than could be produced from other methods.

The person who drew wire first did it with water power around 1350. His name was Rudolf of Nuremberg. Wire was drawn this way for centuries until the steam engine was invented. It wasn't until the 1800s that wire drawing was first accomplished using steam power.

Ichabod Washburn, considered the Father of the Steel Industry, founded a wire mill in Worcester, Massachusetts, in 1831. In the modern age, however, nearly all wire is made by a machine and the process is nearly completely automated. Large pieces of metal, such as iron, steel, copper, aluminum, or other metals called "ingots," are rolled into bars, called "billets." The billets are rolled into smaller rods, which are heated to reduce brittleness, coated with a type of lubricant, and passed through a series of dies that reduce the thickness of the rods even further to produce wire.

To produce wire that is even stronger, the thickness of the rods is increased. When this is done, wire is twisted or braided into rope rather than making a single thick strand. The problem with the thicker strands is that they may break at a single weak spot. Braiding the strands prevents this problem. This is done by twisting wires around a core, which may be wire or rope, to create a strand of medium thickness. Following this, additional strands are wrapped around the core. The result is known as wire rope.

Wire rope was first made in the 1830s. In 1840, a patent for the new wire rope was granted to Robert Newell, an Englishman. The method of braiding or twisting the metal wire grew quite popular and was later used as a strong building material.

John A. Roebling, a leader in suspension bridge construction, pioneered the manufacture of massive steel cables, using the twisting method, for the construction of suspension bridges as well as the Brooklyn Bridge in 1883. Indeed, it is said that there is enough wire rope in the Brooklyn Bridge to stretch all the way to the Moon.

In the late 1800s, another asset of wire became known: it could conduct electricity. The importance of this realization can be seen by reflecting on all the areas electrical wire, in various gauges or thicknesses, is used: buildings, ships, cars, and airplanes. The more power the wire carries, the thicker it must be. If it isn't sized properly, it can heat up and burn the wire out, creating an unsafe situation that is normally handled by a "fuse" blowing or a "circuit breaker" tripping.

Today, almost all electrical wire is made of copper, which has been found to be an excellent conductor of electricity. Electrical wire is almost always encased in some type of rubber or plastic insulating material so that the current running through it won't do any damage. It depends on the type of electrical wire, but many, such as that used in the household, come in a bundle of three wires: a "hot" wire that carries current to the device, a "neutral" wire carries the current back to the source, and a "ground" wire, which is needed in case electricity escapes. This wire routes errant current to the ground—where all electricity seeks to go because of Earth's magnetic pull—rather than letting it hurt someone.

For a while, aluminum wire was used and may still be used in some areas, but it has been implicated in too many house and building fires to be considered safe. The problem is that when aluminum wire is heated up it expands and contracts, thereby loosening the connectors it is secured to, which, in turn, can lead to failure—and fire.

Wire has also been found to be important in a variety of household uses, such as hanging pictures.

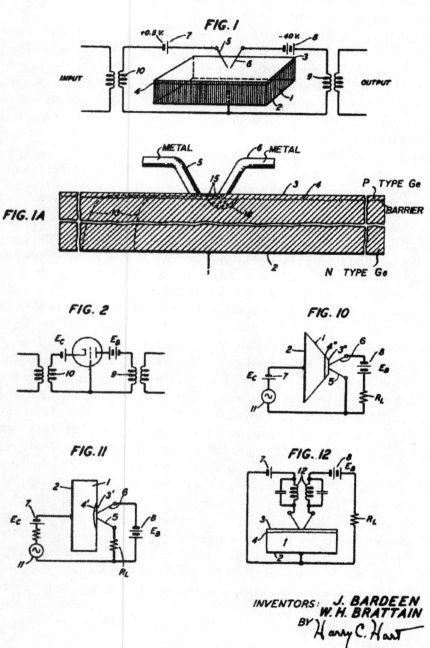

Patent drawing, 1950, by John Bardeen and Walter H. Brattain. *U.S. Patent Office*

Transistor   26

On October 3, 1950, John Bardeen and Walter H. Brattain were awarded a patent for the transistor, but the invention could hardly have come into existence without the involvement of William Shockley, who turned out to be as controversial as he was brilliant.

It all started when a Swedish chemist named Jons Berzelius discovered silicon in 1824, and another chemist, a German named Clemens Alexander Winkler, discovered a substance called germanium in 1886. Both of these substances or elements are known as "semiconductors," because they share a common electrical characteristic: they are considered to have conductive properties somewhere between insulation, which resists electricity absolutely, and metal, which conducts it easily. The amount of material needed is also relatively small.

Bardeen, Brattain, and Shockley did extensive study on these materials, and it resulted in the "transistor," dubbed this name by an electrical engineer named John Robinson Pierce because it could transmit—and amplify—current across a "resistor."

Bardeen, Brattain, and Shockley met at the Bell Labs in Murray Hill, New Jersey, in 1945. Brattain had worked at Bell Labs since 1928, while Shockley joined him in the 1930s and Bardeen in 1945. Shockley had previously known Brattain while in the U.S. Navy during World War II, where they had done work developing antisubmarine systems. Bardeen was a physicist and Brattain had been working in the field of harmonics for seven years when Shockley came aboard.

It was Shockley who first saw the potential for the transistor to replace the vacuum tube. Indeed, in 1939 he had suggested using semiconductors as amplifiers.

AT&T was all ears, as it were. It had a big problem: its patent on the telephone was coming to an end, and this would open the field to a lot of hungry competitors.

The vacuum tube was at the core of the problem. It all started in 1906 when Lee De Forest invented his "triode" in a vacuum tube, which was very good at amplifying signals.

AT&T needed something to amplify signals to send sound and data along telephone lines for great distances—indeed, around the world. To understand this, it's important to know that when electronic signals travel, they do it in stages.

A signal does not go, say, from New York to Malaysia in one shot; rather, it travels from one switchbox in a country, where it is amplified—refreshed like a traveler—to the next switchbox, where it is similarly amplified until it reaches its destination.

AT&T had bought De Forest's patent and it had greatly improved the vacuum tube. Still, there was a major problem: the vacuum tubes that were at the core of amplification were notoriously unreliable. They also used a great deal of power and generated too much heat.

It was felt by AT&T that the semiconductor material might solve these problems. It had conductive qualities, was cheap, and was easier to handle than vacuum tubes.

Using this premise, original experiments were conducted by Brattain and Bardeen that led them to the creation of the "point-contact transistor" in 1947. In 1950, Shockley had developed the "rectifier" and replaced the point-contact transistor with it. This, in turn, led to the development of a device called a "junction transistor" and obviated the need for the point-contact device.

The transistor had a large impact on the size of electrical devices: they could be much smaller, and anyone who lived in the 1950s should remember the advent of the "transistor radio," which was a small, portable radio with high power.

The transistor also changed the electronic innards of the television and had a variety of other applications. While the electronics is complex, perhaps better left to the understanding of a physicist, the key to remember is that the transistor packed a lot of power in a little package—reliably, unlike the vacuum tube.

The work of the three scientists did not go unnoticed. In 1956, they shared the Nobel Prize in physics for their work with the transistor. Shockley departed from Bell Labs in 1956, founding the Shockley Semiconductor Laboratory in what was to become known as "Silicon Valley."

As mentioned, Shockley became quite a controversial figure. His theories of genetics stated that blacks were intellectually inferior to whites. His theories were rebuffed by both the general public and the scientific community. One other thing is remarkable about his life: he died of natural causes.

Steam Engine 27

The steam engine goes back a long way, indeed longer than one might suspect. Hero of Alexandria, a Greek scientist, first spoke of using it to open temple doors. He had built what was essentially a reactor turbine, where water was heated and steam was expelled through two nozzles that turned the turbine and opened the door.

But it was hundreds of years after this, in 1698, that Thomas Savery, an English military engineer, took out a patent on a pump to raise the temperature of water by, as he described it, "the impellent force of fire."

Savery himself had used the work of a Frenchman named Denis Papin to make this primitive "steam engine." Papin was the inventor of the pressure cooker. He was also the first person to realize that water could be drawn up in an enclosed vessel, better known as a "suction pipe."

Savery planned to use the principle in creating a device that could draw water, which was always a problem, out of coal mines. He built a device that consisted of a boiler linked to a pair of containers, valves, and cocks that were worked by hand. The device worked, but among other deficiencies was that it could only take the water up 20 feet or so, hardly a plus with mines being hundreds of feet underground.

It took Thomas Newcomen, also an Englishman, to develop an engine that was to become extremely important in the Industrial Revolution. Newcomen's goal was to develop an engine that could be used to keep water out of the tin mines in Cornwall, which flooded often. The job was being done by employing horses, but it was an expensive solution.

To create the machine that would do it, Newcomen used one of Papin's ideas, which worked along the same lines as the internal combustion engine; that is, steam would be trapped in a cylinder and force the piston to move. Papin had actually built a working model that incorporated this idea, but no one, including Papin, was able to spot its potential.

Newcomen invented an engine that worked, but it had a big problem: it wasted energy. Along with John Calley, a plumber, he developed an engine based on piston action, but in operation a lot of heat was created, which translated into energy. In operation, a boiler would produce steam that would enter a vertical cylinder on top of it. The steam pushed up a rod that rocked a heavy crossbeam that was linked to a pump. The steam was then condensed by a certain amount of

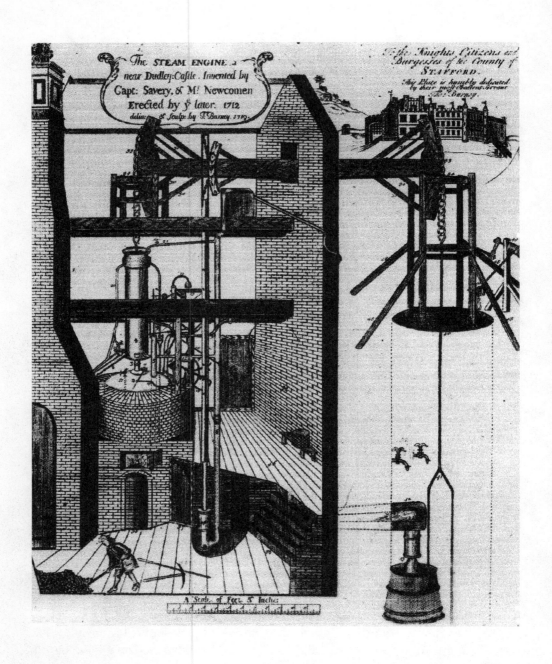

Thomas Newcomen's steam engine, 1712. *New York Public Library Picture Collection*

water entering the cylinder, creating a partial vacuum and allowing atmospheric pressure to drive the piston down again.

The wasted energy came about because the cylinder had to be completely cooled down before it could be heated again. Additionally, air and other gases would sometimes collect in the cylinder and halt the engine's operation. Later, valves were added that increased efficiency and the engine, despite its flaws, was used all over Europe.

The man who perfected the engine, which proved to be crucial in creating power for the Industrial Revolution, was James Watt.

Watt was born in Scotland, the child of a shipbuilder and ship owner. He had tremendous mechanical prowess and eventually opened a shop where he repaired and made instruments.

In 1764, Watt was first introduced to the Newcomen steam engine when he examined a working model brought to him by a client. His analysis of the machine indicated to him that what it needed more than anything else was to increase its efficiency. To do this, he devised an engine that would avoid having to alternately heat up and cool down. The key component was a cylinder that was connected to the engine but in which condensation of the steam could take place without impacting on the engine's performance.

Watt built a model of his machine, unveiling it in 1769 and heralded it as a new way to cut down on the "consumption of steam and fuel to fire engines." And that it did and then some, reducing the amount of fuel needed by 75 percent.

Watt patented his new invention—because it was greatly different than Newcomen's—and it wasn't long before the industrialist Matthew Boulton became interested. Watt, who was not an astute businessman, formed a partnership with Boulton and soon his machine made him wealthy. For his part, Watt worked at continually improving the machine as well as other inventions.

The steam engine continued to be developed during and after Watt's death, and it was to become important in significant ways, both in powering ships and trains. Watt was much honored in his long lifetime and his last name—Watt—was used to describe a unit of electric power.

28

Modern sail boat. *Tom Philbin III*

Sail

The sail's story is eons old and the sailboat remains the earliest known use of wind power. Seafaring boats with sails explored the world and opened trade routes that continue today.

In terms of influence on our destiny, the sail is nothing short of profound, because of the simple fact that men could not row to the same places that sails could take them. Indeed, it took many explorers around the world.

The first sails caught enough wind to push boats through the water and that was all that was needed. Ancient sailors had no idea of what lift and drag meant, nor did they know or care a whit about physics as applied in a nautical situation, but they used such principles every day.

Fiber and hemp were the first materials used for sails, and archaeologists date the use of hemp all the way back to 8000 B.C. in Mesopotamia. The history of the sail is clearly interwoven with the rich history of the hemp plant and its spread from Mesopotamia to Asia, Europe, and Africa. The spread of the seed to the rest of the world was taken by birds and wind.

By the third millennium B.C., hemp had established itself as one of the major fibers of the world. It became an important commercial textile that was traded throughout the world. There is also evidence that the Chinese used it as a superior string to bamboo to equip bows for war and to make scraps of paper, clothing, and slippers, among other uses.

Prior to the sail, hemp was very important in providing a wide variety of string material for sails, ropes, and rigging. These weren't the only uses for it, however. Women of the time were also making hemp sheets that served as clothing.

At the time all this was going on, Venice had become the hemp capital of the world. Indeed, the product was so crucial to Venetian society that the senate decreed that the "security of our galleys and ships and similarly to our sailors and capital [rested on] the manufacture of cordage in our home of Tana." Only the best-quality hemp was used for rope, rigging, and sails on Venetian ships. In fact, hemp helped the Venetian fleet reign over shipping in the Mediterranean until the defeat of Venice by Napoléon Bonaparte in 1797.

Despite the fact that most countries into the first millennia were making use of the hemp plant, it wasn't until the 1500s that it became one of the dominant materials in world society. And it was at this time that the nations of western Europe were struggling to be most dominant.

One of the problems facing sailors at the time was to make rigging and sails that could withstand the journeys across the ocean, facing all kinds of difficult weather that was accompanied by winds that could rip most fabrics to shreds.

But the long fibers of hemp could do it, and so could another product called canefis, of which Holland became the leading supplier. Indeed, it was this material that was used by Christopher Columbus for his sails on his trip to the New World.

This led to the next great leap in sail technology. The Netherlands had the current technology to produce canefis or canvas and suppy it to the West. Still, even as canvas was becoming an important textile for sail making, so important was the demand for hemp that many ships of the time carried hemp seeds in case of a ship wreck so that the sailors could grow crops for needed raw materials (as well as use the seeds for nourishment).

Like many inventions, a number of refinements were made to the sail that would eventually lead to an assortment of sails made from different materials in modern times. Sail making was also affected by other influences. In spite of hemp's many uses, it wasn't a commercially viable crop, so farmers slowly switched to growing other crops. Also, by the 1850s, ships and boats were being powered by steam and even oil. Eventually, other materials, such as nylon, made strong inroads into the sail picture. Nylon was strong, lightweight, easy to manage, and gradually it started to replace other materials for sail making.

Today, sails are no longer used by most developed societies for commerce—their use is now reduced to pleasure. But not in all societies. Many primitive peoples still depend greatly on the sail and use old materials that have been used for thousands of years, including hemp.

29

Native American bow and arrow. *Author*

Bow and Arrow

The bow and arrow is an invention that let the first humans earn their dinner by shooting a grazing or charging animal while they maintained a safe distance. The same relative safety was true when the invention was used as a weapon of war.

An inspired Stone Age man invented the bow and became the most efficient hunter on Earth. It consisted of a slender shaft of wood that could be bent and a cord, sinew, or other flexible material strung tightly between the ends.

The way that a bow and arrow operates is simple, but effective. The cord that attaches the two ends of the bow creates tension from which is propelled an arrow. The arrow is a straight, slender shaft of wood or other material with a sharp point on one end and feathers usually attached to the other end to give it aerodynamic stability.

The bow goes back a very long way. It appears in thirty-thousand-year-old paintings on cave walls in western Europe. Indeed, there is evidence that the bow was used both for hunting and war back to the Paleolithic period.

Materials used for the bow and arrow had a great deal to do with their effectiveness. Various bones and wood were used over time for the bow itself. The material had to meet some qualifications: it had to be readily available to the bow and arrow maker and it had to be strong and flexible.

The arrows were made of various materials. Arrowheads were first made of burned wood, then stone or bones, and finally metal. But it was in 1500 B.C. that the bow took a different turn with the invention of the composite bow, which is made of various materials (wood, sinew, and horns) glued together to greatly increase their natural strength and elasticity. As such, it became the main weapon used by Assyrian charioteers, Mongol horsemen, and English longbow men. At other times, it has been used by massed infantry or cavalry.

Since bows and arrows are relatively easy to make and can produce a relatively rapid rate of fire (when compared with crude firearms after the advent of gunpowder), they were used in warfare long after gunpowder was introduced. They helped people who had a nomadic lifestyle because bows and arrows were relatively small, easy to make, portable, and easy to use effectively.

In the Americas, the bow and arrow made great strides in the Great Basin and the Great Plains. In the prehorse period, both Basin and Plains cultures made highly accurate bows. They needed to be accurate because the American Indians hunted on foot and many times could only take one shot at their quarry.

Great Basin bows were made from local materials such as ash, mountain mahogany, and yew. Bows made from these woods were sometimes backed with sinew to increase their shooting strength and prevent breaking. Bowstrings were made from sinew as well. Although sinew was popular, strings were also made from deer and bear gut, rawhide, plant fiber, and even hair. Most strings were twisted or braided to make them as strong as possible.

The arrows were made from solid woods such as chokecherry, wild rose, and willow. Reed shafts were also used because they were generally light, stiff, and easy to obtain.

Feathers used on the arrows were also highly varied. Practically every bird found in the Great Basin contributed feathers to arrows. The feathers from larger birds, such as geese, eagles, and cranes, were preferred by many American Indians.

In terms of arrowheads, diversity was key and several styles were used in the Great Basin. The arrowheads were made from specific types of available rock in the area. There were at least five types of rocks used. After the arrowhead was shaped, it was secured to the notched shaft with a wrapping of wet sinew or thin rawhide. Most arrowheads ranged from ¾ to 1½ inches in length. Arrows were later designed for specific types of game. Other materials used to make arrowheads included flint, antler, and bone.

Today's bows and arrows are used for sport and hunting. Bows are now made of wood, fiberglass, carbon, and machined aluminum. Arrows are also made of composites and fiberglass. They are lighter and stronger than ever. They also propel an arrow faster and with greater force and accuracy than ever before. The basics, however, haven't changed.

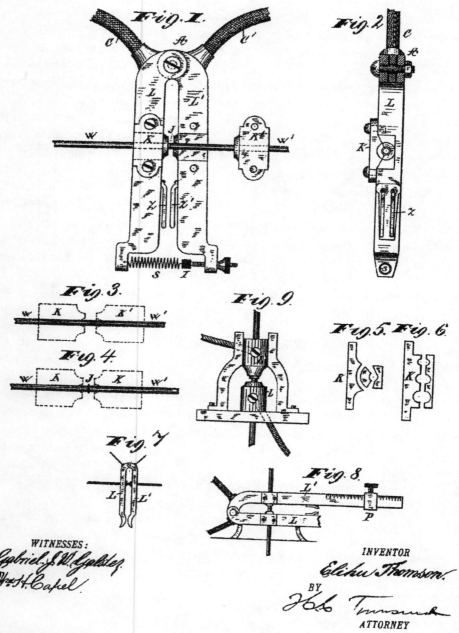

Patent drawing, 1886, by Elihu Thompson. *U.S. Patent Office*

Welding Machine 30

Humans have produced a staggering and impressive array of hand and power tools to repair, maintain, and build the things in their lives, everything from a powered screwdriving machine to a chainsaw that makes quick work of taking a tree down, to a lathe than can create finely turned objects. But there is a quietly efficient, remarkable tool that could make a claim as the greatest tool ever invented: the welding machine.

The welding machine, which has developed various versions of itself over the years, is deeply involved in and important to our lives, and the lives of our forebears. Indeed, almost everything we use in our lives depends on joints being welded, which in its simplest terms could be described as a method to join two pieces of metal permanently and to make them act as a single piece. When necessary, welders also use their equipment to cut through metal (indeed, when there is a collapsed building, you are almost sure to see welders cutting apart the debris to free potential survivors).

Some of the things welders are significantly involved in include the assembly of cars, the making of small appliances, the welding of metal framing on skyscrapers, and the building of ships, bridges, and electronic apparati. Welding machines can function anywhere: inside or outside, as well as under water.

Welding is vital to the economies of many countries, including the United States where it is estimated that it is involved one way or another in 50 percent of the gross national product. It is difficult to imagine how these numbers could change.

The earliest examples of welding can be seen in small gold boxes whose lap joints were joined by welding during the Bronze Age. There is also evidence of its use in the Middle Ages, where one can find items that were welded together by blacksmiths, which was known as "forge welding."

The discovery of "acetylene" occurred in 1836, and the discoverer was a man named Sir Edmund Davy. It was during the late 1800s that "gas welding" and cutting became popular.

Nikolai N. Bernados, a Russian, was the first person granted a patent for welding (in 1885), with another Russian named Stanislaus Olzeswski. Bernados, who was working in France at the time, used the heat of an arc for joining lead plates used in batteries. This was the official beginning of "carbon arc welding."

In 1890, welding took another step forward when an American named C. L.

Coffin was granted a patent for an arc welding process that used an electrode that deposited "filler" in the pieces of metal to be joined.

Around the turn of the twentieth century, arc welding continued to be improved, and some new ways to weld came into use, such as "resistance welding," a process in which the two pieces of metal are joined by passing current between electrodes positioned on opposite sides of the pieces to be welded. This method does not produce an arc. The weld occurs because the metal resists current flow, heating up and resulting in fission and spot welding that usually is done on pieces having an overlapping design. Gas welding was perfected during this period. Various gases were used, and most notably a low-pressure acetylene was developed with a suitable torch. An American named Elihu Thompson invented the "electric arc welder" in 1877 and obtained a patent for it in 1919. World War I saw a tremendous need for producing armaments, and welding, which was quick and could make a joint as strong as possible, was in great demand.

Down through the years, both gas and electric welding continued to be developed and specialized (like everything else), down to the most recent method called "friction welding," which uses rotational speed and pressure to provide heat. This was developed in the Soviet Union.

Laser welding is one of the newest processes, which was originally developed by Bell Labs as a communication tool. But because of the tremendous focus of energy on a small area, the laser turned out to be a powerful heat source, both for cutting and joining.

In sum, no matter what the method or metal (and today just about any metal can be welded) welding is all about heating metals to the point where they become liquid, and when the joints are placed together and dry, they become like the original metal. No joint is stronger.

Cyrus McCormick. *New York Public Library Picture Collection*

McCormick Reaper

The McCormick reaper is named after its inventor, Cyrus McCormick, and while the machine is hardly as obviously dramatic as the automobile, a case could be made that its invention was as important. It had a profound effect on the way people live, was a factor in the Union winning the Civil War, and helped usher in the Industrial Revolution.

Cyrus McCormick was born on February 15, 1809, in Walnut Grove, Virginia, the eldest of eight children. His parents, Robert and Mary Ann McCormick, were both Scotch-Irish and highly religious.

McCormick eventually achieved great success in business, which he ascribed to his healthy physical condition enabling him to spend the large amounts of time that successful businessmen need to invest in projects. He neither smoked nor drank nor took part in any other activity that might have been regarded as sinful at the time. He once described his physical appearance as a young man, "My hair is very dark brown—eyes dark, though not black, complexion fresh and health good. 5 ft. 11½" high, weighing 200 pounds." It was once said that just keeping up with him was a job in itself.

In 1857, at the age of forty-eight, he slowed down enough to marry Nancy Fowler. They were married for twenty-six years and had seven children.

The reaper he invented was designed to cut and store grain more quickly than using current methods. At the time, the operation involved hand-cutting the grain with sharp scythes, and then men and women would pick it up and tie it in sheaves to dry. Using this cutting method, the average man could cut 2 to 3 acres a day. Using the McCormick reaper, the same man could cut 20 acres a day.

McCormick's reaper, which was the same one his father had been working on for twenty years, was a large metal contraption with reciprocating cutting blades that were shielded by metal fingers, a reel to bring the grain against the blade, a divider to isolate the grain to be cut, and a platform onto which the cut grain could fall. It was heavy and horses were required to move it along.

McCormick patented the device in 1834 and started manufacturing it in 1840. He started selling machines in Virginia, but there were problems that made it seem as if the device wouldn't survive. Pulling the reaper exhausted the horses, and it was constantly breaking down. As a matter of fact, for the first few years it was plagued with so many problems that farmers relied on the tried-and-true method of sheer labor to harvest the grain.

Sales of the machine were understandably sluggish until McCormick visited the north central section of America to see how the reaper might perform there. Unlike Virginia, where there were patches of hilly, rocky ground, the north central land was as flat as a billiard table. He figured horses would not be exhausted as quickly when pulling the reaper and the machine would perform much better. In 1847, he moved his operation to Chicago and started manufacturing reapers.

Any good moneymaking idea is sure to be challenged in a variety of ways, and the McCormick reaper was no exception. McCormick had patented his reaper, as already mentioned, in 1834, but a New England man named Obed Hussey had patented his own successful reaper the year before. This was to lead to innumerable court battles and bitterness between the two inventors. Also, McCormick had to go after numerous other companies who infringed on his patent, which usually meant that they would steal the reaper's basic design, add a mechanical flourish here or there, and then patent their reaper as an original design.

Regardless, McCormick was not only a great inventor, he was also an innovator when it came to marketing his product and that was one reason why he became so successful. In 1856, he was selling four thousand reapers annually by use of the then innovative installment plan, so farmers who would not ordinarily be able to pay full price—$100—now could purchase a machine for $35 down in the spring and the other $65 in December.

McCormick also made sure his reapers stayed running. If anything went wrong with a reaper, McCormick or his workers would fix it immediately. He also made sure that farmers who wanted to fix their own reapers could by supplying a how-to repair manual with each machine. In addition, McCormick and his brother made a habit of showing up at harvest time to make sure things were running smoothly.

By the time the Civil War ended, McCormick was reaping huge profits. By then, there were between eighty to ninety thousand machines in use, most on Midwestern farms where the land was flat.

As mentioned earlier, the McCormick reaper helped win the Civil War. For one thing, it enabled northern farmers to harvest more grain for people and horses. It also required fewer men to help with the harvest, thereby more men could serve in the Union army without harming grain production. Two men with one reaper could do as much work as twelve men with scythes and cradles. It also impacted on the Industrial Revolution, allowing more people to leave farms to work in factories.

The McCormick reaper revolutionized farming in America and moved America close to its potential as a farmland.

McCormick died on May 13, 1884. Today, if you drive through the Midwest at harvest time you can still see his basic machine in action, only the name on it will be different: International Harvester.

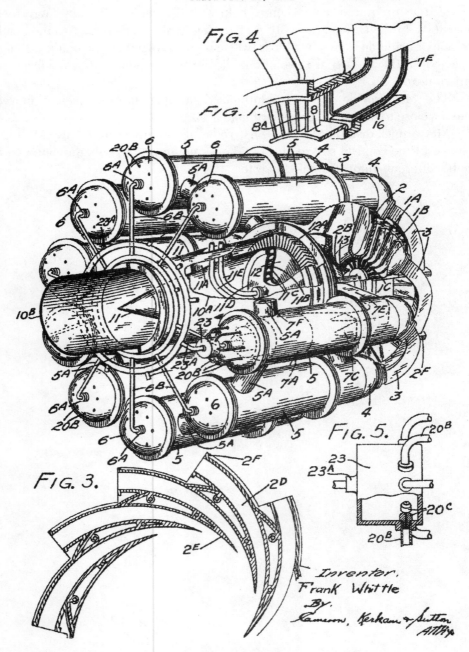

Patent drawing, 1946, by Sir Frank Whittle. *U.S. Patent Office*

Jet Engine 32

One might make a convincing argument for the spontaneity and impatience of youth having partly led to the modern jet engine. While only twenty-two, Sir Frank Whittle, a Royal Air Force pilot and aviation engineer, first began thinking of using a gas turbine engine to power airplanes. During his time (the 1920s), planes were powered by piston engines and propellers, which limited how fast and far they could go. Whittle wanted to go faster—and further.

By 1930, he had designed and patented a jet aircraft engine. Although youth and skill brought him and his engine this far, it would be another eleven years before Whittle's engine successfully powered an aircraft in flight.

Like other significant inventions, however, Whittle was to share his accomplishments with Hans von Chain, another inventor. He started the development of the turbojet engine in the early 1930s while in the midst of his doctoral studies at Goettinger University in Germany. By 1935, he had developed a test engine to demonstrate his ideas.

Both men were engineers, believed in science, and were very mindful of Isaac Newton's third law of physics, that for every action there is an opposite and equal reaction. For example, if you let the air out of a balloon, the balloon will be propelled forward.

In the basic turbojet engine, air comes in at the front of the engine, is compressed, and is then forced into combustion chambers. Here, fuel is sprayed onto it and the mixture is ignited. Rapidly expanding gasses form and push out the rear of the plane. As this occurs, they pass a fan of blades that rotate. This, in turn, is linked to a compressor, which brings in a fresh supply of air.

Over time, a major development was the extra thrust created in the engine by the addition of an afterburner section in which extra fuel is sprayed into the exhausting gases. These hot gases burn the added fuel, thereby adding more thrust. For example, at approximately 400 miles per hour, 1 pound of thrust equals 1 horsepower.

Exhaust gases are also used to rotate a propeller attached to the turbine shaft in turboprop engines for increased power and greater fuel efficiency. Still, jet engines are lighter, more fuel efficient, fuel is cheaper, and the simplicity of their design makes them easier to repair.

The first real run of the first experimental engine was in April 1937. Whittle himself described it as follows:

The experience was frightening. The starting procedure went as planned. By a system of hand signals from me the engine was accelerated to 2,000 rpm by the electric motor. I turned on a pilot fuel jet and ignited it with a hand turned magneto connected to a spark plug with extended electrodes; then I received a "thumbs up" signal from a test fitter looking into the combustion chamber through a small quartz "window." When I started to open the fuel supply valve to the main burner, immediately, with a rising scream, the engine began to accelerate out of control. I promptly shut the control valve, but the uncontrolled acceleration continued. Everyone around took to their heels except me. I was paralyzed with fright and remained rooted to the spot.

The reason for the uncontrolled acceleration was that prior bleeding of fuel lines had created a pool of fuel in the combuster. "The ignition of this was the cause of the 'runaway.' A drain was quickly fitted to ensure that this could not happen again."

During the next year, many development problems were solved and the experimental engine was reconstructed several times. The final version of the engine performed well enough to win the support from the Air Ministry in 1939, something Whittle had sought feverishly.

A flight engine was commissioned to be built. The Gloster Aircraft Company built an experimental airplane. It was completed in March and in May 1941, it took off from the Midlands runway on a historic 17-minute flight.

Chain got his jet engine into the air first, and the results were impressive. Following this, he started to develop the "S-3" engine, which led to the development of the liquid-fuel combuster. Detailed design began in early 1938 on a test aircraft and in early 1939 both the engine and the airframe were completed, but the thrust was below needed requirements. After several internal engine adjustments, the engine was ready to be tested. On August 27, 1939, a test pilot made the first successful flight of a jet-powered aircraft.

Early Great Northern Railway locomotive, built in 1882. *Photofest*

33

Locomotive

The most important development in the creation of the locomotive was the invention of a particular steam engine that when introduced elicited the comment from inventor James Watt that the engine's use should result in its inventor being hanged.

As it happened, the first locomotive was developed by Richard Trevithick for use in hauling raw material from the mines in Scotland. For Watt, safety was the issue, and perhaps jealousy. Watt had invented a steam engine that was only able to withstand 5 pounds of pressure per square inch. Trevithick's engine could withstand 145 pounds per square inch, which translated into the engine having a much smaller cylinder and therefore fewer moving parts.

Trevithick installed his engine in a locomotive, dubbed it the *New Castle*, and ran it on rails leading from a Welsh ironworks, where the work was normally done by horses. The engine proved to be so heavy—though comparatively small—that the rails collapsed under its weight. While a step forward, Trevithick's engine was by no means ideal in powering a locomotive, and it had mechanical problems with the drive and suspension as well as collapsing the rail track due to its weight.

Any invention, of course, is supposed to be better than the device it replaces, but this was easier said than done: the locomotive was supposed to go faster than loaded-down, horse-driven wagons, but in the beginning it didn't.

All of the various factors combined to slow locomotive development until one day a new person arrived on the scene, George Stephenson, whose work ultimately resulted in him being dubbed the Father of the Railways.

Stephenson was born at Wylam, Northumberland, and was the son of a mechanic. He educated himself by attending night school, and by 1802 he had married and his wife had borne him a son.

But tragedy came into his life. His young wife died in 1806, forcing him to bring up his son alone. He was careful to make sure that his son received a full education, unlike himself, and he paid for the boy's education by repairing watches and clocks and cobbling.

Eventually, Stephenson got a job as a mechanic at Killingworth colliery railway, where he earned a reputation as a person who could work wonders with difficult problems. This led to him being chosen by the company to research and design a locomotive for hauling coal out of the mines.

Stephenson's research resulted in the locomotive *Blucher*, which worked quite well. Following his research on the *Blucher*, further research by Stephenson paid big dividends when he discovered a principle that would result in engines being able to go much faster. Namely, that if he fed the waste steam through a narrow pipe through the chimney, the draft in the furnace was greatly enhanced. Known as the "steam blast" technique, it turned out to be the single most important innovation for the locomotive engineer.

But Stephenson did not stop there. He spent the next few years refining and perfecting his invention, including the rails and tracks and making certain his engine was cost efficient. In 1922, he was hired by Stockton and Darlington Railroad, and in 1825 his *Locomotion* went on its maiden voyage, reaching speeds of up to 12 miles per hour. Thereafter, *Locomotion* was primarily used for carrying freight. It wasn't until five years later that it pulled passenger cars on the Liverpool and Manchester Railway.

Liverpool and Manchester was a new railway company, and it was looking for the most suitable locomotive with which to operate. It held a competition, and for this Stephenson built the locomotive *The Rocket*. It left all its competitors in a trail of steam, as it were. It could go 30 miles per hour, a speed faster than a galloping horse.

The formal opening of the new railroad line occurred on September 15, 1830, when, amid cheering crowds, eight brand new Stephenson locomotives were introduced and pulled trains carrying six hundred guests.

Eventually, tunnels had to be built and the land modified to ensure that railway companies could get where they wanted to go. The impact the locomotive had on local communities was great, as it facilitated people going to areas where they had not been before. Not unsurprisingly, word spread to Europe and America, and railroads were built there that modeled themselves after the Liverpool and Manchester railroad. The impact on people in local communities through which railroads could *not* travel was also great. Ultimately, the impact of the locomotive was felt worldwide. Transporting people was not the only thing that the locomotive helped facilitate. Eventually, it also pulled freight trains.

34

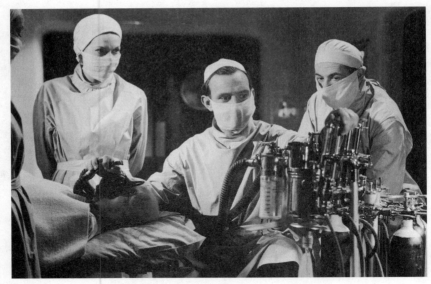

From *Green for Danger* (1947). *Photofest*

Anesthesia

Complex surgical procedures are commonplace in today's practice of medicine. Such operations were, surprisingly, not unknown in ancient times. However, without a means of controlling or eliminating pain, the early Greek, Egyptian, and medieval surgeons lost many a patient to pain-induced shock. Although alcoholic drinks, opiates, and other plant-based drugs were tried, facing surgery was in itself a cause for unbridled fear. Patients were frequently tied down to the operating tables or restrained by assistants as the surgeons wielded their instruments as swiftly as possible, ignoring the agonizing screams of their subjects. It is amazing that a fair number of such operations were actually successful.

Medical technology did not start to catch up with pain until the late eighteenth century. In 1776, Joseph Priestly, a British chemist, isolated the gas nitrous oxide. In 1799, Sir Humphry Davy, experimenting with the gas, noted that in addition to creating a feeling of euphoria—he named it "laughing gas"—nitrous oxide made subjects temporarily insensitive to pain. Davy also put forth the theory that a painkilling drug would best be administered through inhalation, which would allow easier regulation of dosage and withdrawal. Davy advocated its use in surgery, but medical practitioners clung to their old and excruciating procedures.

Michael Faraday, a student of Davy, furthered research on nitrous oxide in the early 1800s and also explored the use of ether, a similar chemical known since 1540. Ether became the subject of not only serious investigation, but like laughing gas was also a popular entertainment among students and enlightened members of society. The "ether frolics" saw people inhaling the gas at parties and enjoying its intoxicating properties. Crawford Williamson Long, a Georgia-born doctor, experimented with ether on himself in the early 1840s. In 1842, Long decided to try ether on a real surgical patient. On March 30, Long's subject—James Venable—inhaled ether from a saturated cloth and had a tumor successfully and painlessly removed from his neck. Long, however, did not make his discovery public. Although he continued using ether in minor operations, he did not try to subject patients to larger doses and did not publish his experimental notes until 1849.

Horace Wells, a Connecticut dentist, began the use of nitrous oxide in his practice in 1844. William Thomas Green Morton, one of Wells' students, continued research with ether. Morton, collaborating with Harvard chemistry professor Charles T. Jackson, was ready to make public his discoveries in 1846. At the Massachusetts General Hospital, Morton assisted surgeon John C. Warren in a successful operation performed in view of a board of prominent physicians. "Anesthesia"—a word invented soon after by Oliver Wendell Holmes—was soon in common use in the United States and Europe. Morton and Jackson were awarded a 5,000-franc prize by the French Academy of Science in 1850. Morton, however, refused to share the award, claiming to be the sole inventor of anesthesiology. Jackson felt likewise and a long and bitter feud between the two ensued for many years. Morton, however, is today considered the Father of Anesthesia.

In addition to Morton's use of ether, other important progress in anesthesiology was made before the turn of the twentieth century. Much work with the use of chloroform, a chemical with similar properties to ether, was done by Scottish gynecologist James Young Simpson. Simpson, an early proponent of ether, began testing with chloroform in 1847. He used it to relieve pain during childbirth, but later found it was more difficult than ether to administer and posed a higher risk to the patient. By the end of the nineteenth century, ether became the preferred anesthesia, and specially trained nurses took to its administration, replacing medical students, interns, and other ad hoc assistants. By 1931, the American Society of Anesthetists was founded and in 1937 the American Board of Anesthesiology began formal certification of specialists. Today, the anesthesiologist is considered of near equal importance to the surgeon in critical operations.

Although this discussion has centered on "general" anesthesia, where the subject is rendered totally unconscious, "local" anesthesia, where an isolated area of the body is numbed, is also widely used and has as long a history. Indians in South America had long known of the effects of the coca plant. Coca was brought to Europe in the mid-nineteenth century and in 1860 German chemist Albert Niemann isolated the chemical agent in the coca leaf and named it cocaine. Sigmund Freud, the noted psychologist, proposed the use of cocaine for eye surgery in 1884. In 1885, William Halstead, a surgeon in Baltimore, used cocaine as a nerve block in minor operations. Unfortunately, cocaine and its derivatives, such as morphine and heroin, are both toxic and addictive. However, related synthetic drugs, such as Novocaine, procaine, xylocaine, and others, are in widespread use in dentistry and certain minor medical procedures. Cocaine and morphine, when properly administered, are still valuable painkillers.

Extending the principal of cocainelike drugs, "nerve-block" anesthesia has resulted in such procedures as the spinal technique. James Leonard Coming suggested blocking the nerve centers in the spinal cord in the 1880s. Such anesthesia renders portions of the body below the area of injection insensitive to pain. Furthermore, the patient can remain awake during the procedure and is less troubled by the afteraffects of general anesthesia. German surgeon August Bier used spinal anesthesia on a patient in 1898. Today, spinal anesthesia is frequently employed for operations in the lower regions of the body and in emergency surgeries where the stomach is full and vomiting could be induced by general anesthesia. Patients undergoing spinal anesthesia are often "up and around"' within hours after the procedure.

Other intravenous agents are also in common usage. Sodium pentothal, a short-acting barbiturate, was developed in 1933 by John Lundy. The drug's chief advantage is its rapid clearing from the bloodstream, which enables pinpoint control of its effects. Curare—used on poison darts by South American Indians—was put to safe use by Harold Griffin, a Canadian anesthesiologist in 1942. Its value is relaxation of muscular groups for surgery in such areas as the abdomen.

Today, modern derivatives have gradually replaced the old standby ether. Halothane, enflurane, methoxyflurane and cyclopropane are now the common anesthetics, although nitrous oxide is still in common use, especially in dentistry. In addition, hypnosis, acupuncture, and electrical stimulation have been deployed with varying success. Computers—with sensors attached to the head—are now being used to control the dosage of anesthesia in response to the activity of the brain.

Although some risks still persist, especially to patients with heart and lung degeneration, anesthesiology has become a reliable and indispensable medical science. Few of our modern-day medical "miracles" could be realized without it. And, in the future new techniques will keep pace with current medical developments.

35

Battery

The first successful experiment that led to the creation of batteries could have been a scene straight out of a science-fiction movie. Working alone in his lab, Italian inventor Luigi Galvani noticed that the leg of a dead frog began to twitch when it came into contact with two different metals. The dramatic but simple conclusion was that there was a connection between electricity and muscle activity. Galvani is associated with today's "galvanic cell" and "volt."

Galvani's timing couldn't have been better. His discovery, among others with electricity, preceded the early development of electrochemical energy storage systems (batteries). Understanding how electricity worked was vital to the development of the battery. Inventor Alessandro Volta made this connection.

The "voltaic pile" (named after its inventor Volta) came along in 1800 and is considered the first battery. But the voltaic pile was just a pile, in this case silver and zinc disks piled on top of one another, but separated by a porous, nonconductive fabric saturated in seawater, which is an excellent conductor of electricity.

The original, seemingly less than scientific design worked. In fact, experiments with different metals and materials continued for sixty years. Still, the voltaic pile remained the only practical form of electricity in the early nineteenth century.

Volta called this pile the "artificial electric organ" (as well as the voltaic pile). When connected to a wire, it supplied electrical currents. Eventually, the connection of electrical energy and chemical reactions was recognized.

Next developed was the "lead-acid battery" by Gaston Plante in 1859. Power was obtained by using lead plates as electrodes that could be charged and recharged.

This application for the battery sparked (as it were) a rush of development on batteries that proceeded quickly. At the end of the nineteenth century, the "dynamo" and the "light bulb" were invented. Due to other industrial developments, there became a need for the storage of electrical energy.

One of the inventors who understood this quite well was Emile Alphonse Fuare, who developed a way to cover both sides of a lead plate with a paste of lead powder and sulphuric acid. This was an important breakthrough that led to the manufacture of lead-acid batteries with cells. Fuare also applied for a patent for making pasted plates for lead-acid batteries.

From this point on, batteries consisted of multiple cells that were electrically connected. These electrochemical cells became the building blocks by which batteries are now connected.

A number of companies popped up that specialized in making batteries. There were also plans for large electricity stores to sell electric power supplies. For instance, William Thompson, later known as Lord Kelvin of Largs (after whom the absolute scale of temperature was named), concocted a plan to supply Buffalo, New York, with electricity from Niagara Falls. Power supplied was to be 80,000 volts, the source a battery with 40,000 cells. Each house in the area was to receive 100 volts of power by tapping groups of 50 cells each. The plan failed for the same reason that the first attempts at industrial manufacture of batteries failed: Faure's cells were not very durable and failed after only a few charge and discharge cycles.

In order to understand how the problem of cell depletion could be solved, classifications of batteries should be understood. There are two major classes of batteries: primary and secondary. Primary batteries, such as the flashlight battery, are used until depleted and then discarded because the chemical reactions that power them are irreversible and after the energy is used up, there is no use for them anymore.

Secondary batteries, such as car batteries, can be charged and reused. They use reversible chemical reactions. By reversing the flow of electricity—that is, putting current in rather than taking it out—the chemical reactions are reversed to restore active material that had been depleted. They are known as rechargeable or storage batteries.

The latest developments have been in the external appearance of batteries. The "wet-cell batteries" were enclosed (to reduce the risk of acid splashing up) and sealed. In the United States today, the "lead storage battery" is commonly used for a variety of purposes, such as in automobiles and construction equipment (where portable, temporary electricity is needed).

Even though the simple definition is suitable for what batteries are and what they do—devices that translate chemical energy into electricity—their variety and usefulness cannot be underrated. Some are small enough to fit on a computer circuit board while others are large enough to power a submarine; some are reusable, some are not. New battery types and significant improvements in the performance of existing batteries have spurred the increased use of batteries throughout society.

36

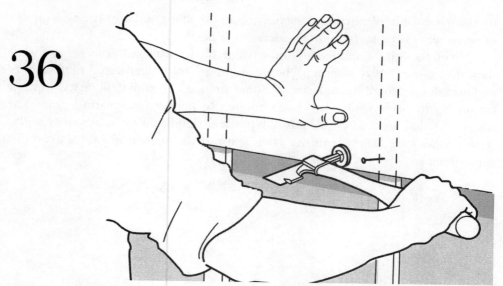

The nail is very simple—but a profoundly important invention. *U.S. Gypsum*

Nail

No one knows for sure when the first sliver of metal was used to join two pieces of wood together or who invented it. Certainly, it was likely to have been a product of one of the metal ages, though the basic idea existed previously of a pointed metal rod with a head in the form of ornamental pins of gold, silver, and copper. There is evidence that the simplest and crudest type of nail has been around since 3,000 B.C.

By the time of the Romans, the hand-forged nail was in use, most notably in the form of the "hobnail," what appears to be a unique Roman device used for joining pieces of leather that remains in use to this day in modern footwear. Nails of many types are fairly common in Roman excavations and shipwrecks dating from A.D. 500. Aside from found artifacts, historical records show that the Romans, in their extensive empire, put the nail to a wide variety of uses as a fastener, including as a particularly cruel component of ancient execution: the crucifixion.

Development of the nail was simple and straightforward, improving only when, like many inventions, advancements were made in the production aspects. Anywhere the earth yielded iron ore people could, with just a rudimentary knowledge of metallurgy (the study of metals), fashion crude "smelters," the heating process that reduces metals to usable forms. From this process, a variety of simple iron utensils, like cooking implements, horseshoes, wagon treads, ship fittings, and the tools to fashion them, could be made, making life just that much easier.

It was for the longest time, up until the 1700s, the job of the blacksmith or more specifically the "nailor" to hammer out nails for society's use in the building of homes, ships, wagons, or storage kegs and barrels as in the cooper's trade. However, it was not all as simple as it seems since the nature of wood varies greatly and will often split along the grain if the shape and dimensions of the nail are wrong. A wide variety of nails were in use by the 1500s including the "brad," or finishing nail, taken from the old Norse word for "spike." This is a very slender, wirelike nail with a cupped head so that the head can be countersunk into the wood and virtually disappear.

The invention of the "slitting mill" in 1565, a mill specifically designed to mass-produce wire and rods capable of being cut into nails, was the beginning of the nail revolution. The slitting mill, usually water powered, put two shafts with big sharp ribs or collars together. When a sheet of metal was passed through the collars, the metal was cut or "slitted" into "nail rods." The width of the rod could be varied by adjusting the setting and size of the collars.

These rods were then farmed out to cottage "nailors" who would then cut the rods into nails with points at one end and heads on the other. A bundle of rods usually weighed 60 pounds and was 4 to 6 feet long. Each bundle was weighed when it was given to the nailor and then the final product was weighed again when delivered to ensure all the metal was accounted for, allowing for some minor waste. The nails were characterized by weight and the number of nails produced by the given amount of metal. The "long thousand" produced about 1,200 nails at 4 pounds and became known as the 4-penny bundle. Larger nails were more profitable because they were easier to make. Produced by the hundreds, they were referred to as "hundred work."

Regardless of these advances, the final shaping of the nail was still done by the nailor, by hand. Between 1790 and 1830, what is referred to as the "Type-A nail" was produced by pounding just one side of the shaft to the point. The "Type-B nail," pounded on multiple sides for a more effective hold, was then developed and used from around 1820 to 1900 when hand-wrought nails were almost completely replaced by precisely made nails produced from machines. However, since restoration projects require the unique look and hold that only a hand-cut nail or piece of hardware can give, there are plenty of nails still produced the old way by artisan nailors, who make their wares often at the site of nail mills hundreds of years old.

The benefits of modern nail making, however, cannot be overlooked. Today, they are available in lengths that vary from 1 to 6 inches. (Any bigger than 6 inches and you have what is commonly referred to as a "spike," as in "railroad spike.") The diameter thickens corresponding to the length. Finishing nails with the small, cupped head are still in use as are an extremely wide variety of nails for specialized uses. One example is the "box nail," especially helpful in instances where the wood is threatening to split. This more slender than average nail is coated in rosin and the friction created by driving the nail with the hammer heats the rosin and bonds it to the surrounding wood. Unfortunately, the box nail has an even greater tendency to bend when driven. Fortunately for the carpenter, however, the Romans had the same problem centuries ago and developed an indispensable companion to the nail: the "claw-hammer," designed specifically to remove bent nails from wood.

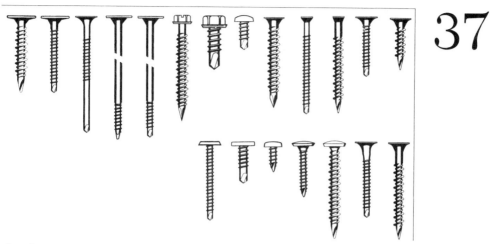

The advantage of the screw over the nail is greater joining strength. *U.S. Gypsum*

Screw

On a visit to Egypt, the Greek philosopher Archimedes is credited with devising a means by which to raise water from one level to another that went beyond a simple waterwheel. The farmers were grateful because they were then able to raise water from the Nile to irrigate their fields on higher planes by using his "water screw."

The design basically involved a spiral made of flexible willow that was soaked in pitch, and wrapped around a wooden cylinder. Boards were then attached to the outside and again soaked in pitch to encase the "screw" in a watertight tube that was reinforced with iron straps. "Pivots," attached at the base, allowed the screw to be turned by hand so that when one end of the machine was submerged in the water source, such as a river or stream, it could raise a lot of water easily for irrigation of nearby farm fields.

While a water wheel was capable of lifting water higher, nothing could beat the volume moved by Archimedes' device, which by approximately A.D. 200 was adapted to other tasks, including being a "bilge pump" that allowed a single sailor to empty water from the bilge of a ship.

By the first century A.D., this principle of the screw was adapted to wooden screws in the wine making and olive oil industries. Indeed, these industries boomed as screw-style wine and olive presses allowed for a more gentle, even pressing of these sensitive fruits, giving the operator previously unheard of control. There were even screw presses designed for fabric, providing well-to-do Romans with neatly pressed togas. Also at this time, a device called a "tap" appeared, which was designed to cut the corresponding internal threads of a nut. Another area the screw principle was applied was in the worm and gear where the worm, basically a screw, propels a gear that connects nonparallel and nonintersecting shafts.

Despite these early applications, the screw with which we are most familiar, the metal fastener, does not show up until the 1400s with, in some cases, its companion the "nut." They were used for holding together pieces of metal, most notably plate armor. At this point, however, the "screwdriver," a tool designed with a blade to fit into a groove on the head of the screw did not exist. Instead, screws had square or hexagonal heads that were driven with wrenches, just as some kinds, such as the "lag screw," are today.

But while no screwdriver has yet been found, there is evidence that something like it was around because screws used to secure armor have been found with slots and notches, indicating that they were turned. By 1744, however, a bit for a carpenter's brace with a flat blade that fit into a corresponding slot on the screw head had been devised. Shortly thereafter, a screwdriver with a handle appeared.

Small wonder that the screw prospered. It was discovered to provide greater strength and a firmer hold than a nail, and when the Industrial Revolution came along there was a tremendous increase in the need for this type of fastener as more and more metal machines were created. Screws for wood—"woodscrews"—also prospered.

Today, the straight-slot screw remains a common sight in both wood- and machine-type screws, never having been fully supplanted by the 1934 invention of the crisscrossed "Phillips head screw." Designed to guide the screwdriver tip to the center of the cylinder, thereby increasing torque and reducing slippage, the Phillips head is the next most commonly used screw. The Phillips, the invention of Henry F. Phillips, a Portland, Oregon, businessman, provides an important advantage in not only one-handed, convenient operation, but power driving as well.

Long before the Phillips head was conceived, an even greater system was devised, one that had a dramatic impact on automotive assembly lines where time, torque, and strength were essential to production. In 1908, Canadian salesman Peter L. Robertson cut his hand one too many times with a straight-slot screwdriver. The enterprising salesman went into his workshop, where he created the "socket head screw" that bears his name. A square bit fits into a corresponding notch in the screw head that almost never strips or slips and maximizes the power of torque that drives the screw home and seldom damages the product. So remarkable was this screw, it came to dominate the manufacturing industry in North America, most notably fastening together the pieces of the Ford Model T and Model A cars. Only very tight patent controls kept the Robertson system from being as universally seen as the Phillips.

38

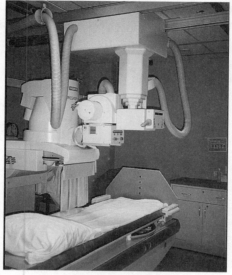

X-ray equipment can be portable or stationary. *Author*

X-Ray Machine

Superman, the popular comic book, movie, and television character, was endowed with many powers, but his "X-ray" vision has been the one that has been most practically adapted by humans. We can't fly (without a plane), leap over tall buildings, or bend steel with our bare hands, but with the use of a device now in common use, we probe the human body, scan suspicious parcels, and benefit from the healing properties of the X ray.

Like radio and television, the discovery of the X ray was linked to the widespread and energetic investigations into electrical phenomena that were conducted during the latter half of the nineteenth century. Sir William Crookes pioneered work on electrical radiation in the late 1870s. His most important contribution, not only in X rays but scientific knowledge in general, was the first cathode-ray (Crookes) tube. Crookes placed two metal electrodes in an glass vacuum tube and fed a high voltage to these opposing plates. He observed darkened spaces near the cathode (negative) electrode and a fluorescent effect at the anode (positive) electrode. He also computed the motion and speed characteristics of the cathode rays within the tube. The Crookes tube is still used today for demonstration and studies in classrooms and laboratories.

German physicist Phillipp Lenard took an important step forward when he built a cathode-ray tube with an aluminum "window" that allowed the rays to disseminate into the air. When Lenard directed the rays at a screen coated with phosphorescent chemicals, he found that the rays caused the screen to glow in relation to the distance and velocity of the rays. He also found that the cathode rays were absorbed by substances in relation to their physical density. In 1892, famed physicist Heinrich Hertz showed that cathode rays can not only be absorbed, but can also actually pass through thin metal foil.

The actual discovery of the X rays and their properties was made—by accident—by William Conrad Rontgen in 1895. Rontgen was born in Prussia and studied and later taught at numerous universities in Germany. By 1894, he had become rector of the University of Wurzburg. Rontgen concentrated his research work on re-creating the experiments and observations of Crookes, Lenard, and Hertz. While he was operating a Crookes tube in a closed, light-proof box, Rontgen observed a flourescent glow in a sample of barium paltinocyanide on a nearby table. This was the first observation of the cathode rays exciting matter at a distance. Rontgen went on to place various objects between the cathode tube and a screen coated with barium platinocyanide. He found that shadows were cast on the screen by the rays in relation to the composition of the subject matter. Lead, for example, absorbed the mysterious rays almost completely, while wood, cardboard, or even aluminum became almost transparent.

"Behind a bound book of about one thousand pages," Rontgen wrote, "I saw the fluorescent screen light up brightly, the printer's ink offering scarcely a noticeable hindrance." Rontgen also found that the rays could form images on photographic plates. He made "photograms" of various metal objects and recorded the bone structure of his wife's hand on December 22, 1895. Rontgen could not account for the exact nature or reason for the propagation of these radiations, so he dubbed them "X rays," a term that readily became popular. Later research showed that the X rays were, in fact, a subpart of the electromagnetic radiation spectrum, similar to radio, ultraviolet, infrared, microwaves, and gamma rays. Being infinitesimally smaller than visible light particles (up to .0000000001 cm—$1/6000$ the wavelength of yellow light!), they penetrate materials opaque to the naked eye.

Rontgen quickly published a complete report on his findings that included detailed categorization of the properties of X rays and the materials that they could effect. Rontgen noted that X rays excited many compounds to fluorescence in addition to barium platinocyanide. Also, he found that the rays could not be focused with normal lenses and traveled in a straight line. Rontgen discovered that platinum, when used as the target material in the cathode tube, yielded more X rays than lighter elements, although any solid body could give off the rays when bombarded by the electron discharge. He was awarded the first Nobel Prize in physics in 1901.

Rontgen's discovery moved quickly from research paper to practical application. X rays were soon put to work in their first—and still foremost—use in medical technology. With the fluoroscope, where the patient was placed between the X-ray tube and a sensitized screen, the bone structure could be clearly seen. Foreign objects, such as bullets, stood out with amazing definition. Many "exploratory" surgeries and agonizing and incorrect "guesswork" were avoided. Photographic plates were used on a larger scale to make permanent records of the X-ray pictures. In addition, the rays were shown to have both healing and harmful effects on the human body. X rays caused hair to fall out on the hands of doctors and technicians using the early devices, and it was found that the rays could destroy diseased tissue and malignant tumors. It was also discovered that X rays could destroy healthy tissue, bone, and blood cells. As a result, lead shields were placed around the X-ray equipment.

During the first quarter of the twentieth century, X-ray research kept pace with its growing use. In the modern X-ray tube, a tungsten wire filament is heated to between 2,300 and 2,400 degrees C, causing the release of electrons. Current then flows between the cathode and anode, with the amount of current being carefully regulated by the operator. The distance between the cathode and anode may vary from 10 millimeters for a low-voltage unit to several feet. The electron beam is then directed at the target, or "focal spot," which emanates the X-ray stream. High voltages are needed for maximum X-ray generation and can range up to 15 million volts! Large amounts of heat can be generated by such currents, so air- or water-cooling devices are used. Tungsten is still used as a target, but copper, molybdenum, and other alloys are sometimes employed. The rays are then directed at the subject, who must be positioned fairly close to the generating tube. An array of voltage controls and circuit breakers is used to regulate the X-ray device, but considerable skill and training is required by the operator to obtain successful, and safe, results.

Radiography, where the X rays are developed on a photographic film, and fluoroscopy, using a sensitive screen, are still being used today much as they were in the 1900s. Radiography has the advantage of producing a permanent record of the examination and subjects the patients or objects being X-rayed to less exposure time, but it does not produce as sharp an image as the fluoroscope. The fluoroscope allows the patient or subject to be observed "in motion," a great advantage in some studies or diagnoses.

The X-ray story, however, does not just have to do with medicine. X rays have been applied to other fields of science and industry as well. Metal structures, both in the factory and in the field, can be examined for faults and structural defects. Computer chips and other microscopic parts and devices are screened by X rays during their intricate manufacture. In the law enforcement and security field, X rays are used to scan suspicious packages and baggage at airports, post offices, and other public places.

Though we haven't become like Superman, X rays have helped us to live better lives and have led the way for some of humanity's most startling discoveries.

39

Military compass. *Author*

Compass

If we lived on Venus, Mars, or perhaps Pluto, we'd have a heck of a time getting from one place to the other accurately if we used the magnetic compass (it depends on the magnetic interior of the planet to work—that is, the needle takes a north-south direction no matter where you are standing) because these planets aren't magnetic. This would also be true on Mercury, Jupiter, Saturn, Uranus, and Neptune.

But the magnetic compass, of course, works on Earth because it has magnetism, which is thought to result from liquid and semimolten cores contacting each other.

Many people think of compasses in an old-fashioned way, such as being used to find one's way across land. But compasses are used to guide ships, planes, and vehicles.

The magnetic compass has been around for hundreds of years, starting with twelfth-century mariners in Europe and China who discovered that when a piece of "lodestone," a magnetic ore, was mounted on a stock in the water it became aligned with the northern polestar. This excited their interest, and experimentation took them to the next step, which was to rub a piece of metal with the lodestone, which also became magnetized, though they hardly knew the word.

To imagine how a compass works, picture Earth as a huge magnet with a north-south orientation that automatically makes other magnetic objects take on the same orientation. This orientation, however, is not quite dead north and south of the axis of the globe. This slight deviance, known as "declination," is not enough to mislead the user for directional purposes, though compasses are commonly made so they are absolutely correct. Of course, declination differences can be hundreds of miles between magnetic and geographic north, depending on where the compass is located when read. In some cases, there are magnetic fields in local terrain, and this can also throw off the north-south orientation. This is known as "deviation."

It was the English who perfected the compass. In past times, wars had been fought mainly at sea, and the fleets had to know which direction they were heading in when there was nothing on the horizon to guide them.

By the thirteenth century, the basic compass had been developed, in which a magnetic needle was mounted in the bottom of a water-filled compass bowl. Only North and South were marked on this primitive compass, but eventually a circular card was cut with thirty other principal points of directions imprinted or painted on it, and this was mounted beneath the needle so directions could be easily read. In the seventeenth century, the needle was made into the shape of a parallelogram, which was easier to mount than a straight needle.

During the fifteenth century—and why it took so long is anyone's guess—people who needed the compass in their daily work started to notice that the compass needle was not fixed at true north. For example, ship navigators noticed this, realizing that the needle actually pointed a little bit off to one side of North. This had to be compensated for when calculating direction.

One of the problems associated with these relatively early compasses was the inability of the needle to hold magnetic north—wherever that might be. Around the middle of the eighteenth century, an Englishman named Godwin Knight invented a way to magnetize steel. This process was applied to the compass needle, which was fashioned into the shape of a bar that was then mounted to the compass. This needle retained its magnetization for long periods of time and made the so-called Knight compass very popular.

During these early days, some compasses, as mentioned earlier, had water in the bowl, but there were some that were dry. There were problems with both: they could easily be knocked awry by a shock and the water (liquid) compass was subject to leakage.

In 1862, the liquid compass became more popular than the dry-card one when the "float" was placed on a card, thus taking the bulk of the weight off the "pivot." Another improvement was the inclusion of a "bellow," which helped keep the amount of liquid in the compass constant. By the beginning of the twentieth century, the dry-card compass was obsolete.

The beginning of the truly modern-day compass started in the 1930s with the development of a housing that was filled with air and protected the needle. This eventually led to the hand-held compass.

Wooden boat under construction. *Author*

Wooden Ships

Great and small wooden ships—sailboats, yachts, and even large sail ships, which are popular for recreation and the highlight of harbor festivals—are today looked on as anachronisms. But the commerce, exploration, and, unfortunately, wars of humanity were carried on from the dawn of existence to the mid-nineteenth century with wooden ships.

Wherever there was water and people, some form of boat was developed. Remnants of dugout canoes and rafts have been unearthed in all parts of the world. The famous Kon-tiki expedition in the 1950s showed that people may have migrated from Asia and Polynesia to South America by using primitive rafts.

Everyone knows how Christopher Columbus crossed the Atlantic on three galleons in 1492, but strong evidence suggests that Leif Erickson with his Viking boats reached North America hundreds of years before. Regardless, wooden ships opened up the New World, and indeed the whole world, for exploration and settlement. Even in Europe, most major cities were situated in ocean, sea, or river ports. The wooden boat was a development that changed and made history.

The first practical and functionally designed wooden craft can be traced to Phoenicia around the time of the fourth millennium B.C. Located in what is now Lebanon and northern Israel, the Phoenicians constructed their "biremes" and "triremes"—boats approaching 200 feet in length and driven, if the wind was right, by a unitary sail. Banks of oars, plied by as many as two hundred rowers seated on up to three levels, powered the craft in calm waters or when added speed was required. The Phoenicians eventually traded and visited all points on the Mediterranean. The famed "royal purple" dye that became a status symbol in Rome was a Phoenician export.

Phoenician timber—the "Cedars of Lebanon"—provided excellent qualities for early shipbuilding, and the Egyptians imported this for their early requirements. An excavation in 1954 near the pyramids at Giza unearthed a nearly intact "burial ship" that was probably used to transport the mummified remains of the pharaoh Cheops from Memphis to the site of the pyramid bearing his name. When restored, the vessel was 145 feet long and 20 feet wide. The boat was constructed of wooden planking held firm by interior ropes. Five pairs of oars—nearly 30 feet long—were used to propel the ship. Pairs of oarsmen were stationed at the bow to steer the rudderless craft.

Ancient Greece succeeded the Phoenicians and Egyptians in shipbuilding and its resultant commercial and military exploitation. Greek merchant vessels were smaller than Phoenician and Egyptian wooden boats. They were often less than 100 feet long and relied on a large, single mast that was mounted midship. A primitive rudder device was mounted astern. The ships also utilized manpower in the form of rowers, seated in open benches.

Greek boats were generally not sailed during harsh weather or at night. The standing rule with Greek seagoers was to beach boats at night or during storms. Greek captains operated conservatively, always keeping land in sight while out on the Mediterranean, although a few ventured out farther as well, into the Atlantic and to other uncharted places. Greek shipwrecks have been explored under the Mediterranean with their cargoes of amphorae—large clay jars—still intact.

The Greeks also put wooden ships to their first major military use. Their "naval" vessels were somewhat longer than the commercial ships, with a 10- to 12-foot "prow" added to the bow. This was used to ram enemy boats. The Greek captains also tried to bring their boat alongside the enemies' and then snap off the enemies' oars and board the ship. The Athenians used larger "biremes" and "triremes" with multiple banks of oarsmen. These rowers were not slaves or conscripts, however, but high-paid mercenaries. Ancient Rome made use of wooden ships as well, but they made no strides in design or application over the Greeks.

During the early centuries of the first millennium A.D., the Vikings refined and to a great extent perfected the smaller wooden ship. The Vikings' Scandinavian homelands—what is most of modern-day Norway—provided the need and the material for ship construction. The numerous fjords—narrow inlets from the ocean—made each Scandinavian settlement favorable to water transportation. And the large Scandinavian forest provided strong timber, and lots of it.

The most important innovation of the Viking ship was the addition of the "keel," a long strip of wood running along the bottom frame of the boat that extended down into the water. This greatly reduced the tendency of the boat to "roll" or "pitch," increased speed, and made steering easier.

Two basic types of Viking ships were constructed: "knorrs" or trading vessels were up to 50 feet long, while the "long ships"—for battle—were as long as 100 feet. The warships had an additional front prow that curved upwards and often featured decorative carvings. Sails were the general power supply, but warships also used up to fifteen oarsmen for rapid approach or withdrawal. Several hundred battle craft were often deployed by the Viking warriors for a major invasion.

The Vikings also developed practical navigational methods by sighting the position of the Sun and the stars. They also were able to determine the first latitude tables by using measuring sticks to record the Sun's midday height. Using these, and other still unknown techniques, the Vikings ventured far afield—or far at sea—making landings and brutal invasions throughout Europe, Greenland, and most probably North America by the 900s.

41

The stethoscope is part of the first line of defense against heart disease. *Author*

Stethoscope

In the early 1800s, a doctor named René Laënnec was working in the Neck Hospital in Paris trying to save hundreds of patients ill with tuberculosis. Laënnec didn't know it, but his work was to lead to one of the great medical discoveries of all time: the stethoscope.

The invention really started one day in 1816. Laënnec was asked to press his ear against a length of wood. Curious, he did and strangely enough he heard the sound of a pin being scratched against the wood. He immediately made a tube from paper to listen to the chests of his patients. He remembered learning when he was a child that sound passes through solids.

To his amazement, the crude invention worked. He could actually hear sounds from inside his patients' chests. The greatest discovery came when Laënnec was able confirm the chest sounds of the dying to what he found at autopsy.

Inspired further, he made a wooden, cylindrical stethoscope and turned it on a lathe. His invention enabled him to propel his reputation in the world of diagnostics: it allowed him to distinguish fatal disease from minor illness and to recommend the need for invasive operations. Not unsurprisingly, Laënnec was belittled and taunted by jealous colleagues.

Over the next few years, he made many alterations to his stethoscope. Eventually, he set up a small shop in his home and began improving his original wooden stethoscope.

The design he came up with consisted of a turned, hollow length of wood. One end of the piece was shaped to fit against an ear and the other was cone shaped. Into this latter end went another piece that had a hollow brass cylinder inside it. The extra piece was used when listening to the heart and removed when listening to lung function.

On March 8, 1817, Laënnec examined forty-year-old Marie-Melanie Basset. This became the first documented use of a stethoscope. And he continued to fine-tune the device for as long as he practiced medicine.

Unfortunately, Laënnec contracted tuberculosis and passed away. Following his death, the stethoscope was widely accepted and soon became the part and parcel of a physician's gear. But his invention was ready to be improved by Pierre Adolphe Piorry in 1828. Piorry installed a second diagnostic device called a "pleximeter" into his design. Also, his invention was only half the size of Laënnec's. It was trumpet shaped, made of wood, and had a removable ivory earpiece and chest piece There was a removable ivory pleximeter attached to the piece. Piorry's design served as the model for most of the "monaural" (one-ear) stethoscopes that were made. Monaural stethoscopes were prized for their small size, but also because they could hold other medical instruments, such as a pleximeter, a percussor, and a thermometer that was stored within the instrument.

The monaural instrument was used exclusively for about thirty years. Even though it was predominantly used into the late nineteenth and early twentieth centuries, it is still being used in the obstetrical field in countries such as the former Soviet Union (the United Kingdom used it as late as the 1980s and may still be using it). However, eventually physicians began wondering if an instrument using both ears would be better than the simple monaural.

In the early 1850s, there was a rush of designs for a new stethoscope that used both ears. This new "binaural" instrument was felt to be the future of auscultation. The first commercially marketed model was that of Dr. Marsh of Cincinnati in 1851. His model contained the first recorded diaphragm chest piece. However, it proved bulky and cumbersome, and quickly faded. The diaphragm would not resurface for about fifty years.

In 1855 in New York, the first stethoscope that closely resembles our modern device was invented. It was binaural, of course, and had ivory earpieces connected to metal tubes that were held together by a hinge joint. Attached to these parts were two tubes covered by round silk that converged into a conical, bell-shaped chest piece.

Today's stethoscopes are extremely sensitive, and it's nothing for them to detect so-called rails in the lungs or abnormalities such as venous hums and systolic clicks in the heart, which can point a person to further diagnostic work and possible treatment.

Like anything else, however, stethoscopes come in varying degrees of quality and efficiency. If doctors are serious about their work, you can be sure that they'll have a top-of-the line model hanging around their neck.

42

Sears Tower, Chicago. *U.S. Gypsum*

Skyscraper

Although many tall structures had been raised throughout history, such as the Egyptian pyramids, the Tower of Babel, the Leaning Tower of Pisa, and numerous church steeples and pagodas, they were constructed of stone and brick. The ever-increasing weight load of the structure had to bear on the lower levels of the foundation. This created massive bases—as in the great pyramids—covering acres of space or "buttresses"—structural "wings"—to take up the stress. These architectural devices severely limited useful space, and, except for tall spires, ultimate height.

All that was to change in the 1800s when a combination of factors made multistoried buildings both feasible and necessary. Populations began shifting to seaports like New York, Boston, and London. Though new jobs were being created by the "Industrial Age," available land for commercial and residential space was limited in the cities. The only way to build was up.

Cast iron sectional construction became the rage because it could closely duplicate the look of fancy brick and stone with lower cost and weight. Even more important, iron girders could be used to create "skeleton" or "cage" construction, which distributed weight more evenly and allowed higher and more rapid building methods.

Fires also played a role in the birth of the skyscraper. The newly crowded tenements and factories in the major cities were most often made of wood, and they burned with great frequency. In 1871, a huge conflagration, set off as legend has it by a cow kicking over a lantern, decimated a large portion of downtown Chicago. The need to rebuild—both quickly and economically—made Chicago, not New York, the birthplace of the "skyscraper."

The edifice that is traditionally known as the first skyscraper is the Home Insurance Building constructed at La Salle and Adams Streets in downtown Chicago in 1884. This ten-story building featured marble outer walls and four large columns of polished granite over the steel framework. It was designed by William Le Baron Jenney, a Massachusetts-born engineer and architect. The building was demolished in 1931. The concept (later summed up in the Bauhaus slogan "form follows function") created by Jenney and other Chicago architects dominated the appearance of most of the city's pioneering buildings. This "Chicago School" included Louis Sullivan and his protégé Frank Lloyd Wright.

New York skyscrapers (incidentally, the term "skyscraper" was not coined in New York or America. It dates back to thirteenth-century Italy where several buildings and towers of heights approaching 300 feet seemed to "scrape the skies.") had origins that predated the Chicago boom. The famed Flatiron Building, a triangular structure that still dominates the intersection of Fifth Avenue and Broadway at 23rd Street, was built in 1902 and was to become the first famous New York skyscraper. Originally known as the Fuller Building, the Flatiron Building (named for its resemblance to the familiar household utensil) is decorated in ornate limestone, and foresaw the "art deco" form that dominated the next phase of skyscraper design.

The first of these imposing landmarks was the Woolworth Building whose fancy Gothic tower rose to a then-incredible 800 feet over lower Broadway in 1913. The Woolworth Building was the be-all and end-all of skyscrapers until the end of the 1920s.

Walter P. Chrysler, the automobile giant, was determined to have a building bearing his name and was equally set on having it be the tallest in the world. In 1928, Chrysler bought property at 42nd Street and Lexington Avenue. Chrysler obtained the services of William van Alen, who had studied architecture at the Pratt Institute, the Beaux-Arts Institute of Design, and the Ecole des Beaux-Arts in Paris.

Alen's plans called for the Chrysler Building to be 925 feet tall. However, he quickly deployed a "vertex" spire that was assembled inside the building and raised into place. With this added spire, the Chrysler Building attained a height of 1,046 feet, the tallest in the world.

Chrysler's justifiable elation in creating the world's tallest structure was not to last long, however. By 1929, work had begun on what is still considered by many to be the most famous building—although no longer the tallest—in the world. The Empire State Building was conceived during the prosperous 1920s by a group of businessmen headed by former New York governor Al Smith. The site chosen—Fifth Avenue between 33rd and 34th Streets—had been formerly occupied by two other landmarks: the Astor family mansion from 1857 to 1893 and the original Waldorf-Astoria Hotel from 1897 until 1929.

The building, planned for 102 stories (1,250 feet in height), went up with great speed, being completed by 1931. "Sidewalk superintendents" marveled at the skilled acrobatic maneuvers of the iron workers and riveters, many of whom were Native American, risking death a quarter of a mile above the street. When completed, the building boasted 67 high-speed elevators, an 86th-floor observation deck, a 102nd-floor windowed observatory, and a mooring mast to accommodate dirigibles! Such landings were never undertaken (wind currents at that height were too risky), but eventually the spire housed transmitting facilities for most of New York's television stations. Opened during the Great Depression, Empire State has attracted millions of sightseers over the years. It was to remain the world's tallest building until the completion of the ill-fated World Trade Center in the 1970s.

Many other tall business structures dotted the New York skyline from the 1930s on. Among the most notable was Rockefeller Center on 6th Avenue between 50th and 53rd Streets. This complex, dominated by the 850-foot Radio Corporation of America (now General Electric) Building, promoted the "city-within-a-city" concept and features sidewalks and streets that are closed to traffic, stores, restaurants, and a central plaza with a public skating rink.

The Great Depression and World War II put the brakes on skyscraper construction both at home and abroad. The end of the war also brought about a change of direction in tall-building design. New, progressive designers advocated the "glass-box" style, using a smaller proportion of available space and creating a light, open, and almost weightless look. The Lever House—at 390 Park Avenue—built in 1952 by the firm Skidmore, Owings and Merrill, was among the first and most famous of these now common buildings. The United Nations Building, a 544-foot structure of such design, was completed in 1953. It is only 72 feet in thickness!

The glass-box style was brought to its epitome with the completion of the World Trade Center. Two 110-story towers dominated the lower Manhattan skyline from just west of Broadway. The complex continued the city-within-a city concept with a large underground shopping mall, directly accessible by subway, an underground parking facility, and a central plaza. Construction of the Twin Towers incorporated many innovations in skyscraper design. They featured a "load-bearing" wall where outer metal columns and beams carried much of the structures' weight, rather than central girders.

But the World Trade Center will be forever remembered for the tragic events of September 11, 2001, when terrorists commandeered two jet airliners that crashed into the towers, causing them to collapse less than two hours later. A major reason the towers came down (besides the damage from the impacts of the jets) was that the intense heat of the burning jet fuel melted and softened the metal structural members, causing each building to literally collapse on itself. It was perhaps due to their construction that the towers stood for as long as they did after the planes had crashed into them (rather than collapsing right away), thus allowing a considerable number of people to escape (although nearly three thousand perished, including hundreds of firefighters, police officers, and emergency personnel).

The Petronas Towers in Kuala Lumpur, Malaysia were completed in 1998 and attain a height of 1,483 feet with 88 stories, and are currently the tallest in the world. Chicago—where skyscrapers were born—holds the number-two spot with its 1,450-foot, 110-floor Sears Tower, completed in 1974.

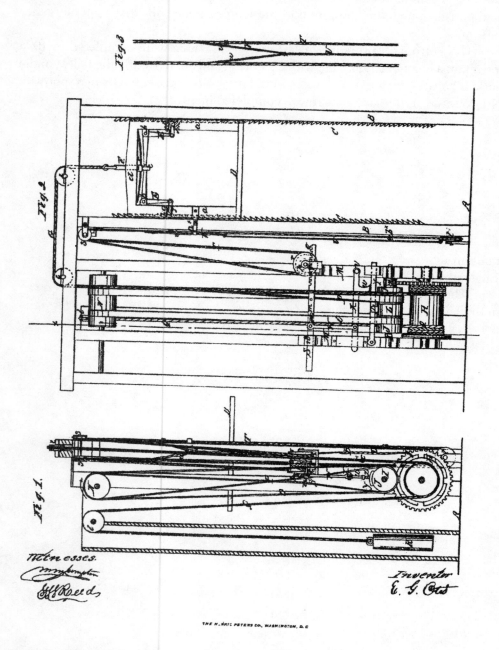

Patent drawing, 1861, by Elisha Graves Otis. *U.S. Patent Office*

Elevator 43

Crude elevators (or "hoists") were used in the Middle Ages and can be traced to the third century B.C. They were operated by animal and human power or by water-driven mechanisms.

In the 1800s, elevators as we are familiar with them started to be developed. Water or steam was used to power them. For example, people would stand in the cab, and then water would fill a hollow tube to the point where the hydraulic pressure would lift the cab into a hollow cylinder.

At first, there was no way to refine the speed of the cab, but as time went by the hydraulic lift was refined and speed regulated with valves of various kinds. Eventually, these "lifts" as they were called were raised or lowered with ropes that ran through pulleys and counterweights. These lifts, which first appeared in England, are the true predecessors of today's elevators.

The first power elevator made its appearance in the mid-nineteenth century in the United States as a simple freight hoist. It operated between just two floors in a New York City building. The inventors, of course, searched for other practical applications for their inventions.

A defining moment in the history of the elevator was when it was proven safe for transporting people. The big moment came in 1853, when Elisha Graves Otis—whose last name still adorns many elevators—designed an elevator that had a key safety feature. If the lifting system of the elevator failed and released the cab, it would automatically stop before crashing into the bottom of the shaft.

The first Otis elevator for passengers was installed in the E. V. Houghwout department store in New York City for the less-than-munificent sum of $300. Otis's elevator was powered by steam. In 1867, Leon Edoux invented and manufactured elevators powered by hydraulic power.

Ten years later, Otis's sons went on to found Otis Brothers and Company in Yonkers, New York. Their company eventually manufactured elevators in the thousands and became a household name in the industry. In fact, by 1873 over two thousand Otis elevators were in use in commercial buildings all over America.

It wasn't until 1884 that the first electric elevator, created by Frank Sprague, was installed in Lawrence, Massachusetts, in a cotton mill. He is also responsible for the push-button control.

Electric elevators were first used commercially in 1889 when two of them were installed in the Demurest Building in New York City. Later, when electricity became more readily available, the electric motor was integrated into elevator technology by German inventor Werner von Siemens. Here, the motor was mounted to the bottom of the elevator and utilized a gearing apparatus to move up or down a shaft with racks in it.

In Baltimore in 1887, electricity was used to power an elevator. This elevator used a drum and a rope to wrap around it. But there was an inherent problem that would ultimately lead to the demise of this type of elevator: the height of the building. As time went by, buildings got taller and taller, necessitating longer and longer ropes and larger and larger drums, until their diameter became impractical.

Motor technology and gears evolved rapidly and in 1889 the direct-connected geared electric elevator allowed for taller buildings to be built. In 1903, the design had developed into the gearless traction electric elevator; such a design could be installed in buildings exceeding one hundred stories. Later still, multispeed motors replaced the original single-speed motors to help with landing and smoother overall operation.

As time went by, ropes were replaced by electromagnetic technology. Various complex signal controls were part of elevators by this time. In addition, safety features were always on the minds of inventors and manufacturers. For example, Charles Otis, son of the original inventor Elisha Graves Otis, developed the "safety" switch that improved the ability to stop the elevator at any speed.

Today, a variety of sophisticated technology is in common use. For example, keypads are used in place of buttons, and switching mechanisms and governors are designed to monitor speed in any situation. In fact, the vast majority of elevators operate independently, and many are equipped with computer technology.

Modern elevators are the crucial element that make it practical to live and work dozens of stories above ground. High-rise cities like New York absolutely depend on elevators. Even in smaller multistory buildings, elevators are essential for making offices and apartments accessible to people with disabilities. Freight and hauling elevators are also indispensable.

Harold Lloyd in *Safety Last* (1923). *Photofest*

Clock

Our daily lives are full of our comings and goings, and all of this has been made very precise. Our lives and our days are planned and timed, all because of the clock.

There is no exact date as to the invention of the first timekeeping device. It is known, however, that timekeeping dates back five to six thousand years ago with the civilizations inhabiting the Middle East and North Africa. The Egyptians in 3500 B.C. had a way of keeping time in the form of obelisks. These were slender, four-sided, tapered monuments that cast shadows on the sand and denoted the passing of time. It was around the same time (3500 B.C.) that the sundial also came into use. This consisted of a round plate with a slanted protrusion coming from its center. As the sun moved, the shadow cast onto the dial would mark the time. Sundials, of course, are still used. In 1500 B.C., the Egyptians were even able to create the first portable sundial, what could be called the grandfather to the watches of today.

Even though both of these are considered to be timekeeping devices, they differ from other forms to come later in that they tell sun time, while our current clocks keep mean solar time. (There are only four times a year when a sundial and a modern clock will have the same time.)

The word "clock" did not come into use until the fourteenth century and its meaning was not the one we know today. It meant "bell" or "alarm." While the first clocks had no internal mechanics, they were capable of duplicating some of the functions of today's clocks, although without the accuracy we enjoy. For instance, the first alarm clock dates back to ancient times. Simplistic in design, it consisted of a candle with lines enscribed into it denoting the passing hours. To "set" the alarm, a nail was pushed into the wax at the appropriate hour. When the candle burned down to the nail, the nail then fell into a tin pan at its base, thus awakening the sleeper.

Water clocks were another way ancient civilizations marked the passage of time. They worked by dripping water into a container, which slowly raised a float that was kept in the container, which in turn carried a pointer to note the hours. The oldest known water clock was found in the tomb of Amenhotep I.

The first mechanical clock with "escapements" did not appear until 1285. The escapement is a mechanism that ticks in a steady rhythm and moves the gears forward in a series of equal jumps. The first public clock to strike the hour was in Milan in about A.D. 1335. Clocks at this point had only one hand, the hour hand, and they failed to keep accurate time.

It was not for another 175 years (1510) that the invention would be improved by Peter Henlien of Nuremberg, Germany, with his creation of the spring-powered clock. While this was a more accurate device, it did have the problem of slowing down when the "mainspring" unwound.

This model was improved in 1525 by Jacob Zech of Prague. He did this by using a "spiral pulley," thus equalizing the pull of the spring. While this had the desired effect of increasing the accuracy of the timepiece, clocks still only operated with an hour hand.

Jost Burgi invented the first clock with a minute hand in 1577. However, it was not until after the invention of the pendulum-regulated clock in 1656 that the minute hand became a practical device.

In the early 1580s Galileo, with his powers of observation and ingenuity, had the inspiration that would produce the first pendulum clock. He found that successive beats of a pendulum always took place in the same amount of time. With this in mind, he and his son Vincenzo began to make drawings and models to find a suitable design. Unfortunately, before they were able to construct the instrument Galileo succumbed to an illness and died. His son, however, did not let his father's vision go unattained and produced a working model in 1649.

Galileo's concept was perfected in 1656 by Christiaan Huygens, who invented the first weight-driven clock through the use of a pendulum. This invention made it possible to keep time accurately, although still only using the hour hand. In 1680, the minute hand finally had its day and it was only a few years later the second hand made its appearance on the timekeeping scene.

In 1889 Siegmund Riefler built a pendulum clock that had accuracy within one hundredth of a second. On its heels came the invention of the double pendulum clock by W. H. Shortt in 1921. This clock operated with one master and one slave pendulum and was accurate to a few milliseconds a day.

While pendulum clocks started to be replaced by quartz clocks in the 1930s and 1940s, they are still in use today. In fact, the old grandfather pendulum clocks are seen as quite the collectible antique.

The quartz clock's operation is based on the piezoelectric property of the crystal. When an electric field is applied to the crystal, it changes shape. Conversely, if one were to squeeze or bend the crystal it would generate an electric field. When coupled with an electronic circuit, the interaction causes the crystal to vibrate, producing a constant frequency signal that can operate the clock's movement. This development was both accurate and inexpensive, thus making it the primary choice of timepieces.

Even though quartz clocks are still used today in great quantities, their precision has been surpassed by the atomic clock.

45

The chronometer enabled ships to find their way across the world's oceans. *Drawing by Lilith Jones*

Chronometer

Today, many people take for granted the knowledge and technology used for safe navigation from one port to another, but only two hundred years ago, lives and cargo were lost at unimaginable rates. Invention of an accurate device known as a "chronometer" changed that.

Because Earth is a ball, all positions, courses, and charts must relate to a sphere. Two coordinates are used to specify position: latitude, an angle set in a plane between the poles, and longitude, an angle set in a plane parallel to the equator. Celestial bodies can be used at sea to calculate latitude positions but not longitude ones. North of the equator, the polestar stays in line with Earth's axis. South of the equator, other fixes are taken.

Calculating longitude was trickier. Europeans got into mathematics and astronomy after A.D. 1000, learning from Arabs, who they came into contact with during the Crusades. Despite the church's suspicion of new knowledge and instruments, advances like the invention of charts and maps accelerated record keeping and exchange of information about the position and features of Earth's spherical surface. Most maps pictured things according to their importance to the mapmaker. Commerce and exploration improved mapmaking from the late 1400s through the 1600s, but maps couldn't be accurate in east-west directions without exact longitudinal determination.

By the end of the sixteenth century, it was understood that longitude was a mathematical problem that dogged European sailing. Distorted maps with guesswork for east-west positioning produced hazardous charting. Christopher Columbus, like all sailors of his time, couldn't calculate his longitude: that's one reason "Indies" and "Indians" are still attached to lands he reached. Eighteenth-century British commodore George Anson lost 1,051 of the 1,939 men who started out with him. The fresh supplies that he needed to obtain from an island near Cape Horn couldn't be found without accurate longitude.

Various governments became interested in the longitude problem.

A British Parliamentary committee consulted scientists like Isaac Newton, who suggested using an accurate seaworthy clock. In 1714, a Parliamentary act offered an award to anyone who came up with something calculating longitude to within a specified degree of accuracy on a test voyage. The invention would have to indicate longitude of the port of arrival in the West Indies—a six weeks' trip at the time.

There were no takers for twenty-three years other than by the Reverends William Whiston and Humphry Ditton, who submitted a plan based on ships moored at specific points along the major trade routes. At midnight, each moored ship would fire a mile-high star shell; the explosion, seen and heard for 85 miles, would inform sailors as to whether the shipboard watches had to be corrected by checking the dead reckoning. The scheme was useless since the moored ships had no accurate clocks.

The revolution in navigation began with the marine chronometer invented by English carpenter and clockmaker John Harrison in the eighteenth century. John and his younger brother James manufactured two clocks that lost no more than one second a month, which were extremely accurate for their era. They decided to make an equally accurate timekeeper that could endure the motion and temperature variation of sea journeys. Their teamwork led to John's success. Their clocks were unaffected by variations of temperature, friction in them was reduced to a minimum, and oil was unnecessary in the mechanism; the timekeeper was similar to earlier wooden regulators but had no pendulum.

The portable "sea clock" (chronometer) could be used for navigational purposes if metal instead of wood were used in many of its parts. They asked the Board of Longitude for financial assistance under the act.

In 1730, Harrison met with Edmond Halley, astronomer royal and board of longitude commissioner. Halley looked over Harrison's plans and recognized that if the clock did work, it was a solution to the longitude problem. He sent him to George Graham, another fellow of the Royal Society, who agreed that if Harrison produced an accurate working clock, the Royal Society would support his approach to the board. Graham himself gave Harrison money to help finance his research and build his clock.

Harrison's first sea clock was completed in 1735. H1 (Harrison's first clock; there were five in all) had no pendulum; it used a balance spring with two 5-pound weights connected by brass arcs. Even when tilted or turned by the movement of the sea, the "regularity of the balances" would not be disturbed. It weighed 72 pounds and was tried out successfully on a barge in England. In 1735, they approached the Board of Longitude again. It agreed to a sea trial. In 1736, H1 sailed to Lisbon.

Harrison went on to design and manufacture H2 with board approval in the form of 500 pounds for development and construction, and with the added caveat that it and future longitude timekeepers he built be surrendered to the Crown. When completed in 1739, H2 was taller, heavier, and not as wide so it took up less deck space. The main innovation in its mechanism and those after it was a "remontoire," a mechanism ensuring that the force on the escapement was constant, which greatly improved the clock's accuracy. H2 was never tried at sea and was the last timekeeper that James was to work on, but John continued the quest.

John Harrison began work on H3, an entirely new design that ended up looking like H2, although it was slightly shorter and lighter with circular balances instead of dumbbells. A bimetallic "curb" to allow for shifts in temperature replaced the "gridirons," but H3 was impossible to adjust without completely dismantling and reassembling it, so Harrison started immediately working on H4, his most famous and important timekeeper. A mere 5¼ inches in diameter, it looked quite different from the earlier timekeepers and was mechanically different, as well. It used oil as a lubricant beneath an especially fine finish that was coupled with wheels and pinions with a high number of teeth to promote mechanical efficiency.

Its trial began in October 1761 when it left Britain for Jamaica. The two-month trip showed H4 to be only 5 seconds slow, a longitudinal error of 1¼ minutes—about the same in miles. Harrison qualified for the award but received no money until George III himself intervened after he saw Harrison's final longitude timekeeper, H5, completed in 1772, which was mechanically very similar to H4.

In the seventeenth century, over two hundred years after the first recorded circumnavigation of the globe, much of the ocean remained unmapped and unexplored. The Pacific stayed unexplored and uncharted until James Cook's three voyages between 1768 and 1779. Cook first left London in 1768. HMS *Endeavour*'s three-year voyage was made without a Harrison chronometer, although its effectiveness at determining longitude had been proven.

But in July 1771, HMS *Resolution* and *Adventure* left Plymouth under Cook's command with Larkum Kendall's copy of Harrison's H4 chronometer on board, enabling Cook to place himself on the surface of the globe with measured latitude and longitude, and to make detailed charts with an accuracy previously unknown. His place in history was fixed.

Small-scale manufacture of chronometers spread. Pierre Le Roy developed one and Thomas Earnshaw produced several (his designs were still used in the twentieth century but without great demand since there wasn't much mass production). Not until the 1850s were three chronometers issued to each British naval ship—if there were only two and they differed, a navigator could not be certain which one was wrong, so the third would provide additional evidence.

46

Early microscope. *New York Public Library Picture Collection*

Microscope

Ironically, it is unclear who invented the first microscope. Like many other inventions, there is a good deal of misinformation about who was first. Some initially believed Galileo made the microscope by reversing his own earlier invention, the telescope. This is disputed, however.

Credit for building first microscope around 1595 is usually given to Zacharias Jansen of Middleburg, Holland. Some say the microscope was inevitable because the Dutch were familiar with magnification in single lenses and then double lenses from which the compound microscope was created.

A common practice at the time was for inventors to make several copies of their invention to give to royalty both as a sign of gratitude and to be inspected. Although none of Jansen's originals given to royalty survived, one of his microscopes did last until the early 1600s. This was long enough for Cornelius Drebbel, Jansen's childhood friend, to examine it and record his observations.

He described the original as being made of three tubes that slid or nested into each other. When fully extended, it measured 18 inches long and 2 inches in diameter. It contained two lenses and diaphragms between the tubes so they slid inside one another easily and cut down on the glare from the lenses. Though unlike today's microscopes, the device did work, having three times the magnification when closed and nine times when extended. It created some excitement at the time.

The next improvement was a three-lens system that was reportedly introduced shortly after Jansen's microscope. It was made possible by using the two-lens Huygens eyepiece system that was then popularly used in telescopes. Within just a few years, word had spread about the microscope and many people began making it and, before long, many learned men of science such as Galileo were using it.

Over time, the compound lens became the most popular model to work on and improve. The feature that made it so appealing to improve was the "achromatic lens," which was used in spectacles and had been developed by Chester Moore Hall in 1729. Although it was difficult to make such lenses small enough for a microscope, work continued on them and by 1900 the achromatic lens had the highest numerical apertures.

Other problems remained as inventors worked on making the lenses more powerful. Contrast-enhancing methods were being sought. This was because the light was always a problem (not enough could be brought around or through the subject being studied) and by adding contrast, one could see the object more clearly.

The stand was also an integral part of the microscope that evolved over time. Originally, the design of the stand was not a stand at all. Instead, the object under scrutiny was impaled on a needle and could be positioned with the turn of a screw. Later, this was replaced with a flat stand because it allowed objects of greater size and shape to be observed. Also, the problem of light was solved. The light source was placed below the stand and filtered up through the object being observed.

Other contrasting methods were developed with greater and greater success until an electronic way of imaging was introduced in 1970. Inventors continued to enhance and influence contrast. The modern-day "electron microscope" can see far beyond what any previous inventors could have even imagined.

Although enormous strides have been made in the evolution of the microscope, modern-day versions work just like the originals. A modern telescope works very much like a refracting telescope but with some minor differences. A telescope must gather large amounts of light from a dim, distant object, so it needs a large objective lens to gather as much light as possible and bring it to a bright focus. Because the objective lens is large, it brings the image of the object to a focus some feet away, which is why telescopes are much larger than microscopes. The eyepiece of the telescope then magnifies that image as it brings it to the observer's eye.

To work well, the microscope must collect light from a well-lit specimen that is up close. The microscope, as a result, does not need a large objective lens. Instead, the objective lens of a microscope is small. To magnify an image, it is magnified by a second lens called an "ocular lens" or eyepiece as it is brought to the observer's eye.

Either way, the microscope is just what its name implies. It allows the user to scope out things smaller than what can be seen by the naked eye.

From *The Story of Esther Costello* (1957). *Photofest*

Braille

When he was just three years old, Louis Braille was working in his father's leather shop when a knife slipped and struck him in the face, resulting in his becoming blind. The loss of sight would have been devastating for anyone, but Braille was not the type to give up. He was a very determined young boy, undefeated by the experience, and was also smart and talented. By the age of ten, he had earned a scholarship at the National Institute for Blind Children in Paris. He was also a musician, having learned to play the organ and cello.

At the school, Braille encountered a system of reading for the blind invented by Valentin Hauy, the founder of the institute. It involved running one's fingertips over embossed letters on paper, but Braille and others found it tedious. The other disadvantage of the system was that it had no capacity to show blind people how to read themselves. It was not "reading, writing and arithmetic," it was just reading.

At the same time, however, there also came along a system that represented a slight advance in reading where one couldn't see, such as in the dark. Called "night writing," it was invented by Captain Charles Barbier, a French soldier who designed it to enable military personnel to read at night. (In those days, it should be remembered, there weren't any convenient portable lights, such as flashlights.) Braille discovered the Barbier system when he was fifteen years old and worked on improving it. Interestingly—and instructive as to the kind of character he had—Braille was of course doing all this when he was blind.

The Barbier system was based on a series of twelve raised dots put down in various positions to represent letters. But Braille invented a system that required only half as many—six dots—and included a series of contractions. For example, the letter *A* was represented by one dot, the letter *B* by two dots, one on top of the other, the letter *C* with two side-by-side dots. Braille's system increased reading speed among the blind. Indeed, they could read twice as fast as when using the Barbier system or about half the speed of a sighted reader.

Braille continued to work on his system, and when he was twenty (1829) the system was published and used informally by the National Institute for the Blind, where Braille had become a teacher. As good as it was, Braille's system, which became known simply as Braille, was not generally accepted by the time he died from tuberculosis in 1852.

Braille's system gradually faded as other systems emerged. In the 1860s, a New York Point system was invented and ten years later an adaptation of Braille's system called American Braille was brought into use by a blind teacher who worked in Boston. But the essential superiority—it was quicker and easier to use—prevailed and Braille's original system became adopted worldwide, including being adopted at an international conference as the official language of the blind in 1932.

As time went by, it started to be used in many countries, and today there is a device—a stylus and accompanying hardware—that enables the blind to write by imprinting the letters on paper. Unlike the writing of the sighted, the blind person writes from right to left.

There are, of course, many books and other documents written in Braille by impressing sheets of paper with zinc plates on which the writing has been imprinted (both sides of the paper may be used without creating problems) and the alphabet is also used by nonblind people for shorthand, musical notation, and in mathematics and science.

It has been discovered that people who go blind late in life have more difficulty than those who are younger at mastering Braille because they are used to other alphabets. To counter this, an Englishman named William Moon invented "Moon type," which like Braille is embossed on paper but is based on modification of letters in the Roman alphabet.

The significance of an alphabet for the blind is staggering. Before Braille, people who were blind were invariably dumped into insane asylums, where they could earn some money by performing crafts that did not require sight. If these people, already inflicted with being blind, weren't mentally disturbed before they entered these institutions, chances are they were when and if they emerged, which was itself unlikely.

Also, of course, it has opened the world of reading to people who would otherwise not be able to, and it has allowed many blind scientists and other people to contribute greatly to humanity.

From *Paratrooper* (1957). Photofest

Radar

People think that during the Battle of Britain it was the British fighter pilots who saved the country by winning the air war with Germany over Britain. As Prime Minster Winston Churchill intoned, "Never have so many owed so much to so few." But there was something else that saved the country and an argument could be made that without it Britain would have lost the war: radar. Like penicillin, which came along just in time to save Allied soldiers from dying from infection as World War II started, so too radar came into being.

Radar was not an idea invented for the war. It had been developed by a number of scientists, but the most important figure was a Scotsman named Robert Watson-Watt, who started working on it in 1915.

Watson-Watt did not work on it with war in mind. Born in Brechin, Scotland, he was originally interested in radio telegraphy, which led him to the London Meteorological Office where he became a research scientist. As planes were coming more and more into use, the concern was that they should be protected from storms and other types of turbulent weather.

He worked on what was fledgling radar—an acronym for radio detection and ranging—and by the early 1920s he joined the National Physical Laboratory's radio section, where he studied and developed navigational equipment and radio beacons.

Radar works on a principle very much like what bats use to avoid hitting each other and objects as they fly around at night in pitch-black caves at high speed. An antenna emits radio waves, and when they hit an object they bounce back as an echo. A measurement can be made of how far away the obstacle or target is by calculating how long the radio wave took to hit it and bounce back.

Eventually, the potential of radar as a military tool was seen, and a number of companies, including some in Germany, strove to develop it quickly. Watson-Watt was appointed to the Air Ministry and the Ministry of Aircraft Production and was given more or less carte blanche to develop radar. By 1935, he had created radar that could detect incoming aircraft 40 miles away. Two years later, Britain had a network of radar stations protecting its coast.

Radar originally had a flaw. The electromagnetic waves were transmitted in a continuous wave that detected that an object was present but not its exact location. Then, in 1936 there was a breakthrough with the development of pulsed radar. Here, the signals were rhythmically intermittent, thus making it possible to measure between the echoes to get a fix on the speed and direction of the target.

In 1939, there was another breakthrough that was tremendously significant: a high-power microwave transmitter was perfected, and its great advantage, which put Britain ahead of the rest of the world, was its accuracy, no matter the weather. It emitted a short beam that could be sharply focused. Another advantage was that it could be picked up by smaller antennae, so radar could be installed on planes and other things.

The practical advantages were many. It allowed the British to deploy their planes with great accuracy in the fight with the German air force, so much so that Germany had to start flying only at night. By this time, the British had installed small microwave units in their planes, thereby enabling fighter pilots to locate and attack German bombers at night. Radar helped in detecting and destroying the fiendish V1 and V2 rockets—the "buzz bombs" that the Germans launched. It was used on D day to locate German defense installations so that attacks could be pinpointed, and was also used in bombing raids over Germany.

Radar, of course, still has many civilian uses. It is invaluable in meteorology, as it can locate dangerous weather events, such as tornadoes and hurricanes. It is also used in all kinds of navigation, including aircraft, boats, rockets, and satellites. More than this, it is used in exploring other planets, including measuring how far apart planets are.

Most people are probably familiar with the use of radar by traffic officers. Using the infamous (to some people) radar gun, traffic officers are able to determine the speed of passing cars with great enough accuracy for the readings to be used as evidence in court. While opprobrious to some, radar in this use has saved untold lives, because drivers who might otherwise put the "pedal to the metal" think twice since a traffic officer with a radar gun might just be around the corner.

Air Conditioning   49

Like many of the world's great inventors, the people responsible for air conditioning didn't set out to change the world. Instead, they were innovators who solved problems they saw around them.

In this case, while simply trying to cool malaria patients at a nineteenth-century hospital in Apalachicola, Florida, Dr. John Gorrie devised a system that blew air over basins of ice that were suspended from the ceiling. It lowered the temperature of the air and made his patients more comfortable. Later on, he created machinery that compressed gas, forced it through radiated coils, and then cooled it by expanding it. The device was patented in 1851 and became the precursor to today's refrigeration systems.

The man who improved air conditioning to the degree that it became practical was Willis Carrier. He is considered the true Father of Air Conditioning.

Carrier's invention also started with a problem. He enjoyed the challenges of problem solving. As an admirer of Henry Ford and Thomas Alva Edison, he shared a common sentiment of his day: with firm conviction and hard work, anything was possible.

Carrier was a driven and disciplined man. His niece recalled one of the last times she saw her uncle as he rested in a lawn chair in his spacious yard with two of his dogs at his feet. "He was sitting there with a notepad and his ever-present slide rule," she says. "I asked 'What are you doing out here, Uncle Willis?' He looked up and said, in all seriousness, 'Trying to figure out the size of a drop of water.' "

Apparently, Carrier always tried to figure things out. When answers didn't come easy, he kept at it. His niece, who lived with him at the time, remembered his persistence. "On one of the first dates with the man who would be my husband, we stayed out rather late," she recalled. "As we pulled into the driveway, the lights in the house were on and I remember saying, 'Uh-oh, Uncle Willis must be up. We'd better go in and face the music.' " When she and her boyfriend approached Carrier, he was working with a notepad and a slide rule. He looked up, lost in thought, and said, "Oh, home early, eh?"

"He actually asked us what time it was," she said. "I think I said 2 but it was really 3. But he didn't know what time it was. We left him in the same position, working, figuring and scribbling, at 9 earlier in the evening. He said, 'Oh, my, it's past my bed time. Good night.' "

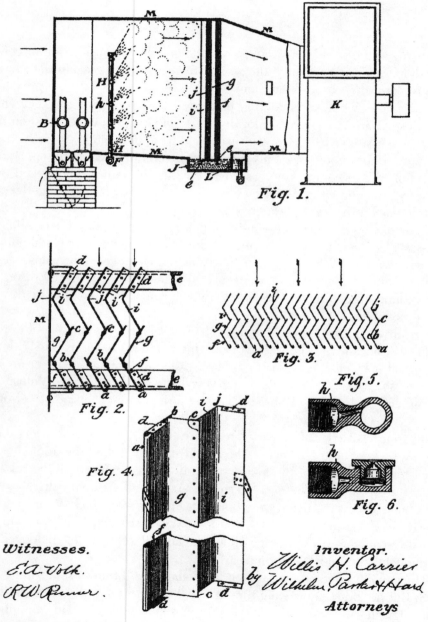

Patent drawing, 1906, by Willis H. Carrier. *U.S. Patent Office*

A year after he graduated from Harvard University, he operated a colored printing machine at the Buffalo Forge Company. He found that the warm temperature in the plant was affecting the final size of the color photos because fluctuations in the heat and humidity had caused the printing paper to alter slightly, enough to make the color prints misalign.

Carrier came to realize that he needed an environment with a stable temperature, and he eventually succeeded in creating one. His system cooled and dried the air inside the plant by blowing it over two sets of coils, one cooled by artisan well water (cold water) and the other by an ammonia refrigerating compressor. The system worked and was reliable for maintaining cool temperatures and the right amount of humidity.

His invention was first installed in 1902 at the Sackett-Wilhelms Lithographing and Publishing Company in Brooklyn. According to Carrier, the machine was the first to perform air conditioning's four functions: to clean the air, cool it, circulate it, and control humidity. Carrier received his patent for the "Apparatus for Treating Air" in 1906, the same year Stuart Cramer, a North Carolina engineer, coined the term "air conditioning."

In 1915, Carrier and six friends founded the Carrier Engineering Company. He continued to refine and develop the air conditioners and by the early 1930s, air conditioning was installed in a variety of commercial buildings.

Ironically, though, Carrier himself didn't believe the family home outside Syracuse, New York, needed air conditioning. His stucco and stone home was a big, beautiful place that was surrounded and shaded by big trees that provided natural cooling.

50

The Brooklyn Bridge. *Tom Philbin III*

Suspension Bridge

Ask anyone you know to name a famous bridge and the odds are that the answer will be the Brooklyn Bridge, the Golden Gate Bridge, or the George Washington Bridge. It is by no means a coincidence that all three of these giant structures are "suspension bridges." The suspension bridge, with its ability to span the largest of rivers and valleys, has truly been a "bridge" to our modern way of life and travel.

The Roman Empire did much to enhance the evolution of bridge design, as it did with other architectural principles, but little was accomplished in the art of bridge building through the Middle Ages and the Renaissance. By the middle of the seventeenth century, things began to change. An engineering school, the Ecole des Fonts et Chaussees, was founded in Paris in 1747 and devoted much of its work to bridge theory. Its students refined the Roman "arch bridge" design to produce less massive structures with the same strength. Such bridges allowed greater leeway in the water for the ever-increasing boat traffic at such places as the Thames in London—the London Bridge.

By the late eighteenth century, the new "trussed arch bridges," now incorporating iron as well as stone, were featuring main spans as long as 200 feet. Timber bridges were still prevalent. In the United States in the first quarter of the nineteenth century, a 340-foot bridge was built over the Schuylkill River in Philadelphia.

America at this time was rapidly expanding, in both size and population. It soon became apparent that rivers, such as the Hudson, Ohio, and Mississippi, would sooner or later have to be bridged. These arteries had to be accommodated by bridges spanning thousands of feet. In addition, the birth of the railroad signaled the demise of rickety wooden bridges, which could not take the weight and vibration of a locomotive and its cars. Even the primitive De Witt Clinton locomotive of 1831 tipped the scales at 3½ tons. By the end of the nineteenth century, locomotives weighed over 70 tons and exceeded speeds of 60 miles per hour. Bridge structures were tested to the maximum. "Iron girder" and "truss bridges"—some as long as 1,500 feet—became common during the mid-nineteenth century, but there were numerous failures and collapses. The suspension principle was waiting in the wings to take center stage in long-span bridge construction.

The suspension bridge supports the central roadway by means of flexible cables anchored at each end of the structure. The cables are strung between high towers that can be spaced farther apart than arch or "cantilever bridges." The steel-wire cables, though appearing frail, have a higher strength-to-weight ratio than solid structures. Therefore, where the longest concrete arch bridge is 1,000 feet, the longest truss bridge is 1,200 feet, and the longest cantilever bridge is 1,800 feet. The Verrazano Narrows Bridge in New York has a main span of 4,260 feet and an overall suspension span of 6,690 feet.

Serious development of the suspension bridge began in 1801 when James Finley used wrought iron chains to support a 70-foot twin-tower bridge at Uniontown, Pennsylvania. In 1826, Thomas Telford designed and built a 580-foot bridge over the Menai Strait in Wales. This bridge featured twin stone towers and wrought iron chain cables to support a wooden deck. The Menai Bridge soon became world-famous, and is still in use, having been extensively rebuilt in 1940.

America took charge of suspension bridge building in the middle of the nineteenth century. In 1849, a 1,010-foot wire suspension bridge was built by Charles Ellet at Wheeling, West Virginia. When the bridge was wrecked during a storm in 1854, John A. Roebling redesigned it. Roebling was to become the most important advocate and designer of the suspension bridge.

Born in Prussia in 1806, Roebling received a degree in civil engineering at Berlin in 1826. He came to America in 1831 and helped establish a farm settlement with other German engineers near Pittsburgh, Pennsylvania. Roebling became a civil engineer for the Pennsylvania State government in 1837. While working on canal projects, he foresaw the need to replace the hemp-based cables then in use on barges and other structures with wire rope. He perfected the manufacture of multistranded wire rope in 1841, and with its immediate acceptance, he began manufacturing it at Trenton, New Jersey, in 1849. He had earlier built the first suspension aqueduct, the Allegheny Bridge (a bridge that transports water) in May of 1845 at the Allegheny River near Pittsburgh. He soon was approached to design and construct other suspension bridges, such as the railroad bridge over Niagara Falls in 1855, and most important, a 1,057-foot bridge spanning the Ohio River near Cincinnati in 1866.

There was at this time a growing need for a bridge to connect lower Manhattan with the then independent city of Brooklyn, New York. Barges and ferries were both becoming woefully inadequate, expensive, and at times dangerous. But a bridge over the East River would have to be thousands of feet long. In 1867, Roebling was appointed chief engineer to build the Brooklyn Bridge.

His design, as accepted, was to be 1,595 feet and would become the world's longest bridge. The project was started in 1869, but unfortunately Roebling died in an accident. Roebling's son, Washington, was able to see the project to its completion, however.

Washington Roebling, like his father, was a trained civil engineer. Born in Saxonburg, Pennsylvania, in 1837, he graduated from Rensselaer Polytechnic Institute in 1857. He worked closely with his father in the development of the wire rope factory and assisted in the aforementioned Allegheny Bridge.

After reaching the rank of colonel in the Union army during the Civil War, Roebling joined his father again for the Ohio River Bridge project. In 1867, he went to Europe to study underwater foundation construction. The method of using compressed air "caissons" to enable workers to excavate and pour bridge foundations became vital in the construction of the Brooklyn Bridge.

Following John Roebling's death in 1869, Washington assumed control of the Brooklyn Bridge project, which took fourteen years to complete. In 1872, he collapsed after spending too many hours in the compressed air caisson; he suffered from the "bends," a condition that was to plague such underwater construction and diving experiments for many years. Washington's health was permanently damaged by the "caisson disease" and he was bedridden during the ultimate completion of the Brooklyn Bridge. He died in 1926.

The Brooklyn Bridge—which is both beautiful and practical—soon carried not only horses and coaches, but also trains, automobiles, and pedestrians. It heralded the rapid development of suspension bridges in the twentieth century. David Steinman developed a "prestressed" twisted sleet wire, which simplified and reduced the cost of wire spinning. This new technique was first incorporated in the Grand Mere Bridge in Quebec, Canada, in 1929. In addition, designers such as Othmar Ammann incorporated the use of stiffening girders on the roadway span itself to reduce the number of radiating wires.

In 1931, Ammann's most famous design—the George Washington Bridge—joined New York City with New Jersey, spanning the Hudson River with a main span of 3,500 feet.

51

Digital thermometer—inserted in the ear. *Duracell*

Thermometer

Today's bulb thermometer is as common as the common cold. Most people recognize it—Mom put it under your tongue (or elsewhere) when you were sick. It is made of glass and has some kind of fluid in it, usually mercury.

At first glance, the thermometer seems to be a simple invention. It works on a simple premise: that a liquid changes its volume relative to its temperature. Liquids take up less space when they are cold and more space when they are warm. How the liquid reacts to differences in temperature speeds up when the liquid is confined to a narrow tube, such as the glass thermometer casing. This is one of the benefits of this design: the liquid, reacting quickly to temperature, is in a tube that has been calibrated and can be read easily.

Early thermometers were dubbed "thermoscopes" and measured different kinds of liquids. Italian inventor Santorio was the first to put a numerical scale on the device. In addition, Galileo invented a crude water thermometer in 1593 that worked to measure temperature variation.

The earliest thermometer that resembles those of today was invented in 1714 by German physicist, Daniel Gabriel Fahrenheit. He was the first inventor to use mercury in his thermometer—a liquid commonly used today. Sealing mercury in a glass thermometer solved many of the problems associated with using water or other liquids. Mercury avoids the freezing and boiling problems associated with water and does not evaporate.

Modern thermometers are calibrated in standard temperature units such as Fahrenheit or Celsius. Prior to the seventeenth century, there was no way to quantify heat. In 1724, the first scale used was called "Fahrenheit," named after Daniel Gabriel Fahrenheit, who also invented the alcohol thermometer in 1709 and the mercury thermometer in 1714.

Later, a "Celsius scale" was invented. This is also referred to as the "centigrade" scale, which literally means "consisting of or divided into 100 degrees." The scale, invented by Swedish astronomer Anders Celsius, has 100 degrees between the freezing and boiling points of water. He devised the centigrade or Celsius scale of temperature in 1742. The term "Celsius" was adopted in 1948 by an international conference on weights and measures.

Another scale was invented in 1848 to measure what its inventor, William Thomas, who later became Baron Kelvin of Largs, called the "ultimate extremes of cold and hot." Thomas developed the idea of absolute temperature, calling it the second law of thermodynamics, and the dynamic theory of heat.

As time passed, inventors of the thermometer followed a similar path most inventors ultimately travel: they looked for different applications or uses for their invention. The thermometer at this time was limited to measuring the temperature of water and gases.

The first medical thermometer used for taking the temperature of a person was invented by Sir Thomas Allbutt in 1867. These thermometers resembled modern thermometers and eventually different ones were developed for taking the temperature of a person by placing them under the tongue, under the armpit, or in the anus.

Theodore Hannes Benzinger, a flight surgeon with the Luftwaffe during World War II, invented the first ear thermometer. In 1984, David Phillips invented the infrared ear thermometer. This allowed the body temperature to be taken very quickly. Jacob Fraden invented the world's best-selling ear thermometer that continues to be popular today.

Calibrating the two most popular scales of measurement for thermometers was a story in and of itself. Although seemingly scientific, the Fahrenheit scale was made arbitrarily. The story goes that its inventor, Daniel Gabriel Fahrenheit, arbitrarily deciding that the freezing and boiling points of water would be separated by 180 degrees, placed a thermometer in freezing water and marked the level of the mercury as 32 degrees. He then stuck the same thermometer in boiling water and marked the level of the mercury as 212 degrees. He then put 180 evenly spaced marks between the freezing and boiling points.

Anders Celsius, also arbitrarily deciding the freezing and boiling points of water, determined that they should be separated by 100 degrees and marked the freezing point of water at 100 degrees. His scale was later inverted, in that the boiling point of water then became 100 degrees C and the freezing point became 0 degrees C.

Liquids aren't the only substances that change their properties when they are heated or cooled. In addition to bulb thermometers, which are good for measuring temperature accurately, bimetallic strip thermometers work on the principle that metal reacts differently when heated and cooled. In fact, different metals expand at different rates as they are heated. These thermometers are often used in ovens.

The electronic thermometer, also known as "thermo resistor," changes its resistance with changes in temperature. It works like a computer, in that circuits measure resistance and convert it to a temperature that is then displayed.

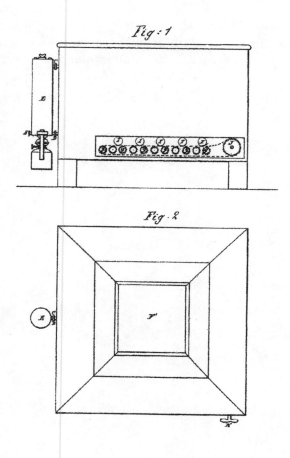

Patent drawing, 1881, by Odile Martin. *U.S. Patent Office*

Incubator 52

Two of my six grandchildren were born prematurely. One was under 3 pounds and the other was under 2 pounds. I remember thinking that they looked a lot like chickens in the supermarket. They were tiny, wrinkled, red little things, all hooked up to tubes in their incubators and required constant monitoring.

I also remember at one point asking one of the nurses, "What would happen to babies like this if they didn't have an incubator?"

"Oh," she said, "they'd die."

In researching this, I found that not everyone would die, but the figures are horrific. In 1888, of all babies who were born premature, some 68 percent died. But the incubator was about to be born and its beginning started, of all places, in the chicken coop.

In 1824, an artificial incubator had been used to hatch chicks for presentation to Princess Victoria in England, where the heat worked wonders in this regard, but none had thought about using it for babies.

In 1878, Stephane Tarnier, a doctor at a children's hospital in Paris, visited the nearby zoo, Jardin d'Acclimation, and saw an apparatus, designed by zookeeper Odile Martin, for the hatching of chicks. It occurred to Tarnier that the device might be used for keeping premature babies warm.

Tarnier had a box designed by a farmer and in 1883 presented his design to the esteemed British medical journal *The Lancet*. Responding positively, it published an article on Tarnier's box complete with the drawings used in the patent. Though the box could still be used to hatch chicken eggs, *The Lancet* noted that it would be "applicable" to other purposes, one of which, of course, was for babies.

Consisting of double walls and a glass top, the incubator was heated by warm water, which could be heated in various ways, that filled the walls. It could accommodate two infants who were withdrawn from the side of the box. The temperature was maintained at 30 degrees C.

The incubator was used at the Paris Maternity Hospital and the death rate among babies who were born 2 kilograms (about 4.25 pounds) or less dropped from 66 percent to 38 percent. Not a panacea, but definitely a beginning. And this was without any special unit being established. In 1893, this was done by Pierre Budin, one of Tarnier's colleagues.

The French emphatically wanted to share the wealth of what they were doing, so in 1896 they sent six of their incubators to the Berlin Exposition. Martin Couney, the assistant of the exposition, asked for and got from a nearby hospital six premature babies to be placed in the incubators. He figured that there wasn't any risk to the babies involved, since they weren't going to survive anyway. But he was wrong about them dying. All six of the babies survived, and the incubator was a great success.

The next year, the same experiment was set up at a British exposition, but no British parents would risk having their children in a French invention. Indeed, in order to fill the incubators, premature babies from France had to be imported.

In so many words, the incubator's popularity was at a standstill. In 1897, *The Lancet* commented that incubators were not yet in "general use in England."

One complaint about these early incubators was that they didn't have automatic temperature controls, thus requiring constant monitoring by a nurse or some other person to ensure that the incubator did not get too hot or cold. Others complained that incubators were only being used for rich parents' babies.

But progress was being made, particularly in the area of temperature control. One incubator at a county fair was displayed that had a bimetallic strip of the kind used in home heating thermostats that would serve to adjust the temperature automatically.

There was also some concern about displaying the babies at the expositions. For example, *The Lancet,* in its February 1898 issue, asked, "Is it in keeping with the dignity of science that incubators and living babies should be exhibited between the aunt-sallies, the merry-go-rounds, the five legged mule, the wild animals, the clowns, penny-pee shows, and amidst the glare and noise of a vulgar fair?"

Still, there was a positive side: The incubators received free publicity and were brought to the attention of the world at large. As the saying goes, there's no such thing as bad publicity. Ninety some odd years after the 1896 Berlin Exposition, I for one am glad that they got it.

53

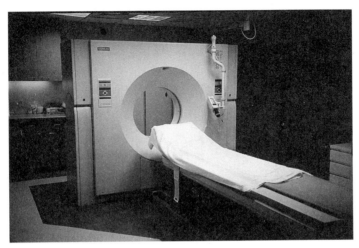

Modern CAT scanner. *Author*

CT Scan

The "computed tomography" (CT) scan allowed doctors for the first time to see the soft tissues inside the body. It was a giant leap forward from the conventional X ray that preceded it. The X ray allowed only the outline of bones and organs to be seen.

The CT scan, sometimes known as a "computed axial tomography" (CAT) scan, uses a computer to produce detailed cross-sectional images of the body. In this way, doctors can examine the body one narrow "slice" at a time to pinpoint specific areas of concern.

CT was invented in 1972 by British engineer Godfrey Hounsfield of EMI Laboratories in England and independently by South African–born physicist Allan Cormack of Tufts University.

The first CT scanners were installed between 1974 and 1976. The original systems were designed to scan the head of the body only. Whole body systems with studies became available in 1976.

The first CT scanner developed by Hounsfield took several hours to acquire the raw data for a single scan or slice and took days to reconstruct a single image from this raw data. CT scans of today can collect up to four slices of data images from millions of data points in less than a second. In the latest scans, an entire chest can be scanned in 5 to 10 seconds using the most advanced multislice system.

During the history of CT, many of the improvements have been made in patient comfort, more anatomy being scanned in less time, and an increase in image quality. Much of the recent research has been made to provide excellent image quality for diagnostic confidence and the lowest possible X-ray dose. In this way, the patient has been offered the highest-quality image possible while keeping the exposure risk at a minimum.

CT is based on the X-ray principal. As X rays pass through the body, they are absorbed at different levels, thus creating a matrix or profile of X-ray beams of different strength. The X-ray profile is registered on film, creating an image. For CT images, the film is replaced by a banana-shaped detector that measures the X-ray profile.

From the outside, a CT scanner looks like a large square. The opening where the patient enters is 24 to 28 inches in diameter. Inside the machine is a rotating frame with an X-ray tube attached on one side and the banana-shaped detector mounted on the opposite side. A fan beam of X-ray is created as the rotating frame spins the X-ray tube and detector around the patient. Each time the X-ray tube and detector make a 360-degree rotation, an image or slice is focused to a specific thickness using lead shutters in front of the X-ray tube and detector.

As the X-ray tube and detector make the full rotation, the detector takes numerous snapshots of the X-ray beam. During the course of one full rotation, about a thousand profiles are sampled. Each profile is subdivided spatially by the detectors, fed into about seven hundred individual channels, and reconstructed backward by a dedicated computer into a two-dimensional image of the slice that was scanned. In order to control the entire scan, multiple computers are used. Much like a dance with multiple partners, the "host" computer leads the way and orchestrates the operation of the entire system.

Included in the system is the dedicated computer that reconstructs the "raw CT data" into an image. A workstation allows a technician to monitor the exam. The CT computers have multiple microprocessors that control the rotation of the "gantry," movement of the table, and other functions, such as turning the X-ray beam on and off.

A crucial difference between CT scanning and X rays is that CT enables direct imaging and differentiation of soft tissue structures, such as the liver, lung tissue, and fat. It is especially useful in searching for lesions, tumors, and metastasis. For helping in the diagnosis of such material, CT reveals the presence, size, spatial location, and extent of the material. Another example of the benefit of CT scans is the imaging of the head and brain to detect tumors, blood clots, blood vessel defects, enlarged ventricles, and other abnormalities, such as those of the nerves or muscles of the eye.

Due to short scan times, CT can be used for all anatomic regions. Some of these include bony structures such as vertebral discs, complex joints like the shoulder or hip as a functional unit, and fractures, especially those affecting the spine.

The benefits of CT scans can be seen in trauma centers. The reasons are that CT scans are fast and simple, they enable a quick overview of possible life-threatening pathology, and they help doctors and surgeons make rapid, yet accurate, decisions on the course of medical treatment. With the advent of spiral CT, the continuous acquisition of complete CT volumes can be used for diagnosis of blood vessels with CT angiography.

54

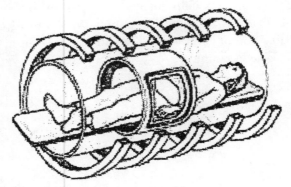

MRI scanner. *National Institute of Health*

MRI

Magnetic resonance imaging (MRI) was once called nuclear magnetic resonance imaging, but the word "nuclear" was dropped about fifteen years ago because of fears that people would think there was something radioactive involved, which there is not. The MRI is a way of getting pictures of different parts of the body without the use of X rays or computed tomography (CT) scans and has a number of advantages.

Like many high-tech inventions, the MRI had a long gestation. Indeed, it took years of research prior to July 3, 1977, when the first MRI was performed on a human being. This event scarcely made a ripple outside the medical community, but it led to greater numbers of machines being produced every year.

It took almost 5 hours to produce the first image. It was, by today's standards, quite primitive. Still, Raymond Damadian, a physician and scientist, along with Larry Minkoff and Michael Goldsmith, labored long and hard to perfect it. Indeed, they named their original machine *Indomitable*.

As late as 1982, there were just a handful of MRI scanners in the entire United States. But today there are thousands. Images that used to take hours now take seconds. The machines are generally smaller—though they hardly seem small when you see them—have more options, and are quieter than earlier models, but the technology itself remains complicated.

An MRI looks like a giant cube. It's about 7 feet tall, 7 feet wide, and 10 feet long. There is a horizontal tube running through the machine from front to back. This tube is called the bore of the magnet, which is itself the key component of the device. While lying on his or her back, the patient is slid into the bore on a special table. The type of exam being performed dictates whether the patient goes in head first or feet first and how far.

The MRI works like this: A radio wave antenna is used to send signals to the body and then receive them back. These returning signals are converted into pictures by a computer attached to the scanner. Pictures can be taken of any part of the body and can focus on thin or thick slices of tissue.

Working in tandem with radio waves, an MRI can select a point inside a patient's body and essentially ask it to identify its tissue type. The scanner is capable of great precision: the point or section chosen might be a half-millimeter cube. The MRI system then goes through the patient's body, point by point, building up a map of tissue types, and then integrates this information in the form of two- or three-dimensional pictures. The images are much better when compared with any other imaging tool, such as X rays or CT scans.

Another advantage of the MRI is that it can "look" at the soft tissues of the body. The brain, spinal cord, and nerves in particular are seen much more clearly with an MRI than with regular X rays and CT scans. Also, because muscles, ligaments, and tendons are seen quite well, MRI scans are used to look at knees and shoulders following injuries. Safety is another advantage of the MRI. Unlike the radiation of an X ray or the limiting benefits of a CT scan, an MRI poses little health risk and is safe for the majority of patients.

Some disadvantages are that some people may be claustrophobic (fear of closed spaces) or may experience some anxiety associated with the hammering noises that are emitted while the scan is taking place. Earplugs are offered to patients for this reason. One disadvantage for the technicians—not the patient—is that the environment of the scanning room must be carefully controlled. The magnetic force inside the machine and around it is very strong and anything that is metal can be sucked into the machine. For this reason, the technicians must make sure there are no metal items on the person's body (metal parts within the body are generally not a problem) or in the examining room. This includes credit cards, film, or any other device with a bar code that can be damaged by the effects of the magnetic force.

The future of the MRI seems unlimited. It has been in widespread use for less than twenty years and already it has made gigantic medical contributions.

55

Drywall factory. *U.S. Gypsum*

Drywall

The miracle of drywall can only truly be appreciated by knowing that prior to its invention in the late 1890s, walls were plastered and this was an art that was very difficult. Indeed, only a very few highly specialized craftspeople could be said to do it well enough to create the flawless plane of wall texture commonly and easily achieved by the humble drywall hanger.

Plaster is a mixture of the naturally occurring gypsum (hydrous calcium sulfate), water, lime, and, depending on the use for which it is intended, sand, cement, or some other material. Its use for structure interiors has been found in Egyptian pyramids dating back to 2000 B.C., but as an element of building, it can be found as far back as 6000 B.C. in Anatolia (modern-day Turkey).

The mineral itself appears in either a powder or rock form that boasts two particles of water for every particle of calcium sulfate. When the rock is ground and then heated, a good portion of the water is released. Reintroducing water and adding lime to the dry mix of gypsum makes it quite plastic and shapeable for usually about 10 to 15 minutes before it dries into a relatively soft stonelike form.

Because gypsum is plentiful in most places on Earth, the discovery of these properties was fairly universal. However, the product of the process of reducing it to powder by heat came to be known as "plaster of Paris" from the plasterers in the city who found and worked with huge deposits under the Montmartre District of the French capital. Plastering, as previously mentioned, is not easy. To plaster a wall, "laths" were mounted on the studs of the soon-to-be wall. In the old days, the laths were slats of wood nailed horizontally to the studs with a little space maintained between each lath to allow the plaster to grab hold of the surface. Two more layers of plaster were applied subsequently, another "scratch" coat, and then the last layer called a "putty" coat to create the final surface, which was completely dependent on the hand that held the tool to smooth it. This process was known as "wetwall."

In the late 1800s, Augustine Sackett and Fred L. Kane came up with the idea to make a new kind of construction material: premade boards of straw paper and tar pitch. Unfortunately (or perhaps fortunately, depending on your point of view), the pitch bled through every wall covering that was applied. Not the kind to give up so easily, Sackett and Kane switched to manila paper instead of straw and plaster of Paris instead of pitch. When it dried, the result was phenomenal: A tough yet smooth board that could be applied directly to the wall studs and hold any kind of surface covering. And it was (and is still is) easy to use.

Sheets are usually stacked horizontally against studs and are secured with special drywall screws. The edges of the drywall are tapered so that when the joints are sealed with "joint tape" and "spackling"—three coats—it is hard to notice where the joint is, that is if done by a competent spackler. The screws are also spackled to hide their heads.

The material is easy to cut as well. When scored with a sharp knife on one side, pressure can be placed against the cut line to break the drywall, also known as plasterboard, into neat sections.

Drywall comes in various lengths (up to 12 feet) and thicknesses. It is also available in so-called waterproof form for use in bathrooms and other areas where water can be a problem.

With all this, drywall is also one of the cheapest products available because the raw material used to make it is very cheap.

Surprisingly, when drywall first appeared it did not catch on immediately. It took ten years of hard work to promote the product with builders, but once it was seen how sharply the Sackett Plaster Board reduced the time it took to put walls in, it caught on rapidly. By 1909, Sackett and Kane were producing 47 million square meters of Sackett Plaster Board a year. That same year, Sackett sold the company to U.S. Gypsum, where it was redesigned to make it lighter and stronger. By 1917, U.S. Gypsum gave the material the common name it uses to this day: Sheetrock.

Drywall is an important invention because without it homes and other buildings would take a great deal longer and cost a lot more to build.

Admittedly, a plaster or wetwall job is still superior to drywall. If you try to drive your fist through a drywall wall, you will probably succeed. Throw the same punch at a plaster wall and you will end up getting treatment for a broken hand.

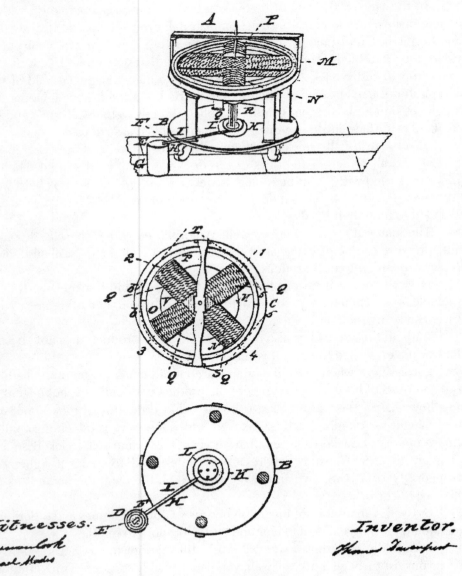

Patent drawing of DC electric motor, 1837, by Thomas Davenport. *U.S. Patent Office*

Electric Motor   56

As a youngster, Michael Faraday worked as an errand boy for a book binding shop in London. Born in 1791 to a poor family, he was an extremely curious child and questioned everything. His urgent need to know motivated him to read every book he found and he vowed to write a book of his own.

Faraday was a scientist, inquisitive by nature, and this curiosity led him into an exploration of things mechanical and electromechanical, particularly force. But he always felt that the scientist was suppressed or in some way put down so that all of the ideas that he had would never emerge. As he put it once: "The world little knows how many of the thoughts and theories which have passed through the mind of the scientific investigator have been crushed in silence and secrecy by his own severe criticism and adverse examination; that in the most successful instances not a tenth of the suggestions, the hopes, the wishes, the preliminary conclusions have been realized."

Still, in 1831 he succeeded in building the first electric motor. Joseph Henry was working on the same kind of motor at the same time and is also given credit for the invention. In 1837, more improvements were made by inventors but it wasn't until 1887 that Nikola Tesla introduced the alternating current (AC) motor. All other motors up until that time had been using direct current (DC). The DC electrical motor had been invented by Thomas Davenport, an American blacksmith.

The difference between DC and AC is important to understand because most motors today use AC. DC is best explained by batteries. Batteries are filled with electrolyte fluids with two different pieces of metal in them. These metal pieces have different electrical properties, in that one end of the battery is negative and the other is positive. Electricity goes round and round directly and in the same direction.

AC, on the other hand, involves current that goes back and forth in a different direction when a magnetic field is applied to it. That is, when a magnet is turned 180 degrees near the current of electrons, the electrons flow in the opposite direction. But when the magnet is rotated quickly, the electrons flow back and forth in alternating waves.

Faraday's initial, previous successes with two devices led to the electric motor, in what he called "continuous electromagnetic rotation," that is, a continuous circular motion from the circular magnetic force around a wire. It wasn't until ten years later, in 1831, that he began his famous experiments in which he discovered electromagnetic induction. His experiments formed the basis of modern electromagnetic technology.

Also in 1831, he made one of his most significant discoveries: the "induction" or "generation" of electricity in a wire by means of the electromagnetic effect of a current in another wire. To determine this, he used his "induction ring." This induction ring was considered the first electric transformer.

Later on, Faraday completed another round of experiments and discovered magnetoelectric induction. He achieved this quite cleverly. First, he started by attaching two wires through a sliding contact to a copper disk. By rotating the disk through the poles of a horseshoe magnet, he noticed a continuous DC. This came to be known as the first crude generator. Directly from these experiments came the modern electric motor, generator, and transformer.

Faraday was also the discoverer of electromagnetic induction, electromagnetic rotations, the magneto-optical effect, diamagnetism, and field theory.

The electric motor resembles a cylinder with a metal casing around the outside of it. On the inside are spools of wrapped wires and a magnet. The AC turns the shaft that, when connected to other things, powers a host of other machines. Put simply, the electric motor is designed to convert electrical energy to mechanical energy. It takes electricity and turns it into energy that can be used by us.

To put Faraday's invention in perspective, consider where the electric motor is now (although most people would say they don't see an electric motor everyday, like they do other inventions, they still depend on and benefit from it). Electric motors are inside many household devices we use everyday. They come in all shapes and sizes and work tirelessly to power clothes washers, dishwashers, alternators on cars, and many other devices.

Although many people don't recognize what it does and how it works, the electric motor has become a very useful invention in modern times.

Barbed Wire   57

There is no doubt that whether you are using barbed wire to keep something in or out, it is quite effective.

The making of barbed wire and its use in fencing dates all the way back to A.D. 400. It started as "unbarbed" wire. Hot iron was pulled through dies to produce short lengths of smooth wire, which was available in various gauges. This technique was improved over the centuries, and in 1870 it was possible to buy good-quality wire in a myriad of sizes and lengths. While often used in fencing, however, it did not deter the animals the way ranchers or farmers wanted. And often crops would be eaten or trampled by marauding livestock.

Ranchers tried many different types of fence construction: wire drawn between treated wood posts or cement, wooden rails, stone, and even thorny shrubs. Nothing really worked.

Joseph Farwell Glidden, a teacher from New York bought a farm in Dekalb, Illinois, in 1843. A few months later he was at a local fair and saw his first example of a barbed "deterrent." It consisted of a wooden rail with sharp nails at intervals hung inside a wire fence. This sparked his creativity and he set about to do better.

The problem for Glidden was to create a product that would jab livestock to the point of deterring it, while at the same time not injuring it severely. He settled on barbs that could smart, but could not really maim. In 1874, Glidden obtained his first patent on his invention.

But a few problems developed from his invention. When livestock encountered it for the first time, they were not wise enough to proceed slowly, which resulted in some injuries.

Some groups were against it too. For example, people who wanted ranges unfettered by fencing saw that the use of barbed wire would mean the end of their livelihood. Trail drivers feared that their herds would be blocked from Kansas markets due to all the fences that were being installed by settlers. Religious groups even went so far as to call it "the work of the Devil" or "Devils rope."

To stop its installation, some people vandalized barbed wire by cutting it and sometimes even battled with the owners. Eventually, these actions led to laws being passed that made wire cutting a felony. These laws successfully stopped the cutting.

Even though this fighting was making people question the use of barbed wire, neither Glidden or Hiram B. Scutt, who had formed companies, suffered.

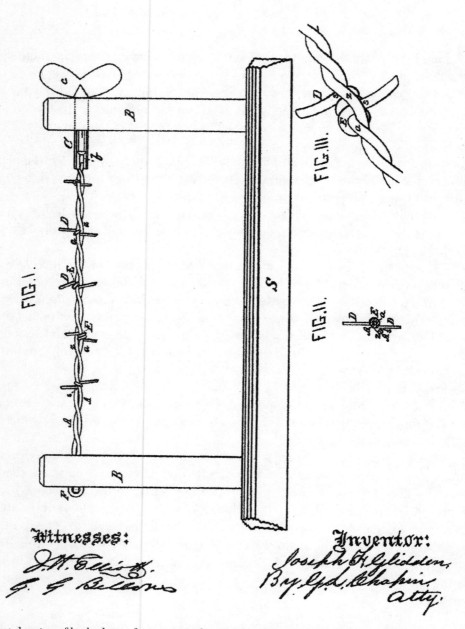

Patent drawing of barbed wire fences, 1874, by Joseph Farwell Glidden. *U.S. Patent Office*

While people feared its use, it was necessary to protect livestock and crops. It was a good idea that could not be denied.

To win back public opinion, at one point John Gates, a traveling barbed wire salesman, challenged a group of Texas ranchers to enclose a group of their wild-range steers in a barbed wire corral in the middle of San Antonio. It retained them without injuring them. Gate's stunt was a grand success, helping to sway public opinion; it also resulted in the sale of hundreds of miles of barbed wire.

While it may look the same, barbed wire is constructed from a variety of metals and in different designs. Indeed, during Glidden's lifetime 570 barbed wire patents were filed. In fact, so many people tried to jump on the bandwagon that there was a three-year legal battle over who really had the rights to the patents.

In the end, Glidden was declared the winner. This obtained him the title of the Father of Barbed Wire. This decision forced many of the smaller companies selling patent rights to merge with the larger steel and wire companies.

Today, barbed wire is still used in the containment of livestock and crops. Its uses, however, have changed and grown with society and its fears. No longer used to simply contain livestock, it is also used to contain people. It is uncoiled around the perimeters of jails, military facilities, and even warehouses and buildings. Two of its more historic and notorious uses were in no-man's-land during World War I and in the German concentration camps.

There are also new forms of it that are extraordinarily nasty, such as "concertina wire." One sees it in deep, high, and seemingly endless roils around prisons. For obvious reasons, it works far better than stone.

58 Condom

While it has often been the subject of risqué humor, a dirty little secret, the condom has nevertheless been a simple yet profoundly important invention that humans have devised not only to avoid unwanted pregnancies, but also to prevent the spreading of disease.

It is difficult to say when the first condom was invented. The earliest illustration of a condom in use was found in Egypt and goes back more than three thousand years. It is not easy to tell what use the person wearing the condom had in mind. He may have worn it for sexual or ritualistic reasons, perhaps even to prevent disease. The Romans, some historians say, were also familiar with condoms, ghoulishly making them from the muscle tissue of soldiers they defeated in battle. In the 1700s, the rapacious Casanova is said to have worn condoms made of linen, although their permeability could not have prevented anything—conception or disease.

Amazingly, some condoms have survived for hundreds of years. The oldest discovered, made of animal and fish intestines, were found at Dudley Castle near Birmingham, England. The theory goes that they were used to prevent the transmission of sexual diseases during the war between the armies of King Charles I and Oliver Cromwell in the 1600s.

Rubber condoms were gradually developed during the 1800s and after the invention of the vulcanization of rubber by Charles Goodyear in 1899 they were mass-produced.

While rubber condoms were used for a long time, the technology gradually recessed as condoms made from animal intestines were manufactured in the 1940s and 1950s to be used more than once. After use, they were washed, coated with petroleum jelly, and stored away for use in the future. Condoms made from animal intestines are still used, but are not designed for reuse.

Condoms were not always as highly regarded by health professionals as they are today. Around the turn of the twentieth century, an organization called the American Social Hygiene Organization advocated men not using condoms. It stated that if venereal disease (VD) was an issue, the user would be subject to catching it whether he chose to have sex with or without a condom. In other words, he would pay the consequences of having sex no matter what type of precautions he took.

There were other important people opposed to condom use as well, and during World War I they were heard loud and clear. A number of military leaders, including no less than the secretary of the navy, were opposed to their use and characterized condoms as immoral and un-Christian.

The result was harrowing. During the war, only the American forces were without the use of condoms, and it was estimated that over 70 percent of all military personnel required their use. As a result, aside from illegitimate births that resulted, more soldiers from the United States contracted sexual diseases than soldiers from any other country.

Not all important people opposed their use for military personnel. Indeed, a young assistant secretary of the navy named Franklin D. Roosevelt had the gumption to order, when the secretary was away from the office, that all sailors be issued them.

Margaret Sanger, the founder of Planned Parenthood and arguably the most important person ever in promoting birth control, saw a double standard during her lifetime regarding condom use. She observed men being issued condoms to protect themselves against sexual diseases, but women could not get them to give to sexual partners and protect themselves against unwanted pregnancies. Indeed, during Sanger's lifetime many women, without an effective method of birth control, had many children, which led not only to economic hardship, but also to physical peril, with many women dying in childbirth. Indeed, Sanger believed that the real reason her mother died at the age of forty was not because of the tuberculosis she had contracted but because of the eleven children she had given birth to during her childbearing years.

America was not the only country with short-sighted views of condoms. Nazi Germany, obsessed with creating a master race, forbid the use of condoms at home so that their world would be peopled with strong, blond-haired, blue-eyed Aryans. However, their concern did not extend to soldiers fighting far away from their German wives and girlfriends. They were issued condoms to protect them against disease: it's hard to fight a war when you're also fighting VD.

As World War II dawned, many military leaders adopted a different view of condoms. They knew that soldiers might well carry VD back with them from foreign lands and infect their wives and girlfriends on the home front, so they openly encouraged condom use. The United States, for example, had many training films warning about VD and the use of condoms; sometimes, they used slogans that were quite graphic, and even smutty. "Don't forget, put it on before you put it in."

The condom almost became history in the 1960s. It was the time of sexual revolution, where young people openly had sex in a context of "free love" and with other birth control methods, such as the Pill and intrauterine devices.

But the development of AIDS changed that because scientists quickly determined that when a condom was used it was rare that the essential transfer of bodily fluids, necessary to introduce HIV infection, would take place.

Today, condoms are still going strong, and there are many different types, sizes, and colors. But the fundamental principle that it came to be known for hasn't changed: preventing sperm from going inside a female.

The telescope is a window to the universe. *Author*

59

Telescope

Many people probably think of the telescope as a device that can be used to bring various everyday objects—and people—optically closer than the human eye allows. It does, of course, but it has evolved into a device that can see the planets—and beyond.

Ironically, although the telescope was improved and made known by Galileo and other scientists, the invention was actually the product of craftsmen, and its history is partially a mystery due in large part to the fact that most craftsmen at the time were illiterate; they simply weren't able—or perhaps willing—to document its development.

The parts of the telescope—the concave and convex lenses—have been available since ancient times. But it wasn't until glass of reasonable quality became available in the major glass making centers of Venice and Florence that it was considered useful. The lenses used in telescopes had other purposes as well, such as being used as hand-held magnifying glasses, which people were using in lieu of eyeglasses by the thirteenth century.

It was then that craftsmen began to make small, rounded, polished convex lenses that were installed in frames. Eventually, around 1350 the first eyeglasses came into existence and indeed became a symbol of learning: You wouldn't wear eyeglasses unless you could read.

It is estimated that by about 1450 the lenses (both convex and concave) and mirrors needed for making a telescope were available, but no one manufactured them. Indeed, it's a question for historians. Some believe the lenses and mirrors of the correct strengths weren't available until later.

It is speculated that as early as the 1570s Leonard Digges and Thomas Digges had actually made a "telescope" consisting of a convex lens and a mirror in England, but it seems to have been an experimental device that never caught on for the world in general.

Instead, the telescope was introduced in October 1608 in the Netherlands. The Dutch government had actually considered issuing a patent to Hans Lipperhey of Middleburg and then to Jacob Metius of Alkmaar on a device for "seeing faraway things as though nearby." But something stopped them.

There was nothing wrong with the device. It was quite simple and it seemed to work. It was made of convex and concave lenses placed within a tube, and the combination magnified three or four times. A remarkable thing happened to their application for a patent: the Dutch government found the device too simplistic to award a patent but instead awarded the two inventors money to build several binocular versions. News of the invention spread rapidly across Europe and by April 1609 3-powered spyglasses could be bought in spectacle maker's shops in Paris.

The first dramatic use of the telescope was by Thomas Harriot, who observed the Moon with a 6-powered instrument in August 1609. However, it wasn't until Galileo made his presentations that the invention's fame increased.

Galileo built and presented an 8-powered telescope to the Venetian senate in August 1609 and used a 20-powered instrument to look at the sky later that year. With his instrument, he saw the Moon, the satellites of Jupiter, and several other stars more clearly. He published *Sidereus Nuncius* in March, 1610.

Beginning in the 1640s, the length of telescopes began to increase and the lenses became more sophisticated and powerful. Indeed, during Galileo's lifetime the power of the telescope reached about thirty times magnification.

By 1704, Isaac Newton had invented a new kind of telescope. Instead of glass lenses, a curved mirror was used to gather light and reflect it back to a point of focus. This reflective mirror acted like a light-collecting bucket. The idea was that the bigger the bucket, the more light it could collect. The "reflector telescope," as it became known, made it possible to magnify objects millions of times. This kind of mirror is huge (236 inches in diameter) and is used today at the Special Astrophysical Observatory in Russia, which opened in 1974.

The latest and greatest telescope is the Hubble Space Telescope. Originally designed in 1974 and finally launched in 1990, the Hubble orbits 375 miles above Earth, continually photographs and peers into space, and sends data back to scientists all over the world. It uses pointing position, powerful optics, and state-of-the-art instruments to provide stunning views of the universe that cannot be made from ground-based telescopes.

Thanks to on-orbit service calls by space shuttle astronauts, Hubble continues to be a state-of-the-art space telescope. Hubble is the first scientific mission of any kind that is specifically designed for routine servicing by space-walking astronauts.

60

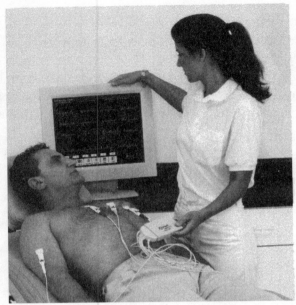

One of the most important tools for preventing and treating heart disease. *Midmark*

EKG Machine

Have you ever wondered what makes your heart beat, hour after hour, day after day? The answer lies is in a select group of heart or "pacemaker" cells that can generate electrical activity on their own.

Located in one specific chamber of the heart, these pacemaker cells allow charged particles to enter through their plasma membrane. These charged particles activate the pacemaker cells, thereby causing the heart to contract. This, in turn, produces a predictable pattern of activity in the heart that is measured by a electrocardiogram (EKG) machine; if the pattern does not follow the accepted norms for the heart, a physician can see it right away.

To look a little closer at how this heart activity takes place, the pacemaker cells are situated in the right atrium of the heart, which is one of the top two heart chambers, and these cells travel to the left atrium, causing right and left atria. After a slight delay, allowing the atria to contract, the two lower chambers of the heart or ventricles fill up with blood. The signal then travels to what is known as the left and right bundle branches, which in turn cause the ventricles to contract, pushing out the blood.

All of this electrical activity produces waves that, as mentioned earlier, are measured by an EKG. The EKG monitors three distinct parts of heart beats.

They are the "P wave," as electrical activity spreads over the atria, the "QRS complex," when electrical activity spreads over the ventricles, and the "T wave," which is the recovery phase of the ventricles.

The EKG machine of today is, like other inventions, the result of an evolutionary process of development and refinement. The first device developed was known as a "galvanometer" in 1794. This device sensed electricity in the human heart rather than measured its current. It wasn't until 1849 that the existing device was enhanced by Emil Du Bois-Reymond. The electrical current could then be measured by adding a two-position switch. This device was called a "rheotome" (flowslicer).

In 1868, Julius Bernstein, a student of Du Bois-Reymond, improved the rheotome by allowing the time between stimulation and sampling to be varied. Known as the "differential rheotome," it was the first device to measure EKGs. At this time, most of the EKGs were of frog hearts and the electrodes were placed directly on the heart itself.

Since more sensitivity was needed to accurately measure the electrical impulses of the heart, the "capillary electrometer" was invented. It was devised by Gabriel Lippman in 1872. Still, another problem remained: The electrical activity of the heart could not be accurately measured without opening up the chest cavity.

Augustus Desiré Waller was the first to discover this and successfully record the electrical activity of the human heart in 1887. He called his initial report of the procedure an "electrogram." He later named it "cardiogram." Still later, the term was introduced that we use today: "electrocardiogram."

Willem Einthoven began developing his own galvanometer in 1900. His work was more sophisticated than Du Bois-Reymond's had been more than a century before. Einthoven's labors were motivated by his not liking the capillary electrometer. Instead, he created the "string galvanometer," which was introduced in 1903.

His EKG was initially manufactured in Germany by Edelmann and Sons of Munich. The manufacturer who later built it was Cambridge Scientific Instrument Company.

The first EKG machine that appeared in the United States was an Edelmann String Electrocardiograph brought by Alfred Cohn in 1909. The first EKG machine manufactured in the United States was designed by Professor Horatio Williams and built in 1914 by Charles Hindle. Cohn received the first Hindle EKG machine in May 1915.

An interesting and dramatic fact at the time was that on May 20, 1915, Cohn's machine demonstrated that the patient getting the EKG was in the process of having a coronary.

During the evolution of the string galvanometer, its size decreased from 600 pounds in 1903 to 30 pounds in 1928. Another improvement was that the size of the electrodes was reduced.

In 1920, Cohn introduced the strap-on electrode in the United States. Ten years later, the Cambridge Scientific Instrument Company of New York introduced the German silver direct-contact plate electrodes. A suction electrode was developed by Rudolph Burger in 1932 for the precordial leads. These were later modified into a suction cup, which has been standard for quite a while.

The next part of the development of the EKG machine was the use of vacuum tubes for amplification. The first of its type was developed in the United States by General Electric. Once the cathode-ray tube was introduced, the physical characteristics of the recorder on the EKG machine improved.

The next step was the introduction of the amplifier-type EKGs that led to the development of the direct-writing instruments. These instruments were able to translate electric impulses to ink marks on a page and a continuous illustration of a patient's heart activity was then available.

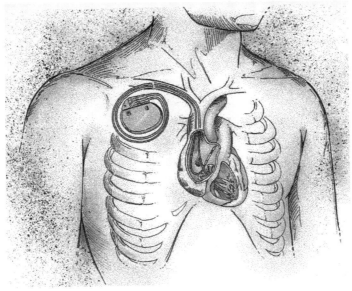

Diagram of pacemaker installation. *Medtronic*

Pacemaker

The heart has to be exact in all of its functions. Each beat has to beat properly, as even a minor problem can cause death. But various problems can be detected by an electrocardiogram machine, and if necessary the heart can be regulated by a pacemaker.

One of the major problems is arrhythmia, where the heart does not beat as it should. There are two types of arrhythmias: tachycardia (heart beats at a rate faster than normal) and bradycardia (heart beats at a rate slower than normal). Treatment for tachycardia consists of cardioversion (delivery of a broad depolarizing shock to a restricted region of the heart). Rapid bursts of pacemaker impulses that are properly timed and placed can often stop the tachycardia. An implanted pacemaker can also restore the lower heart rate to a more physiological value that will improve cardiovascular function as it has for millions of people.

Fibrillation is another major problem that affects the heart. This is the uncontrolled beating of different parts of the heart. Ventricular fibrillation is a fatal arrhythmia of the heart in which the victim will die in minutes if it is not corrected. Atrial fibrillation is a less serious arrhythmia because the ventricles are still pumping. However, it can lead to problems if it is not corrected. Heart block is another problem caused by the interruption of the internal electroconduction system of the heart. These are a few heart problems that people encounter that can be conquered with help from pacemakers, defibrillators, and modern technology.

Cardiac research is not new: electricity was used for the stimulation of the heart in the late eighteenth and early nineteenth centuries. Albert S. Hyman is believed to be the founder of the artificial pacemaker, but he probably was not the first. Mark C. Lidwill, an Australian physician, and physicist Major Edgar Booth demonstrated their portable pacemaking unit in 1931: an apparatus with one pole applied to the skin and another in the appropriate cardiac chamber.

Credit for the pacemaker is given to Paul Maurice Zoll. Zoll had studied the relationship of alcoholism to heart disease during medical school. He eventually became a cardiologist and worked with Dwight Harken, a fellow Harvard graduate who successfully removed foreign bodies from the heart. The irritability of the heart while being operated on impressed Zoll.

Back in Boston, Zoll resumed his research in 1945 and in the late 1940s met a sixty-year-old woman with Stokes-Adams disease (a conduction blockage). When she died, Zoll, aware of earlier work in the 1930s demonstrating the use of electrical stimulation of rabbits and dogs, decided it should be possible to stimulate the heart. He borrowed a stimulator from Otto Krayer and in 1950 used an esophageal wire to restart a dog's heart. Later, he elicited the same response with an external stimulus applied to a human chest. In 1952, he applied this treatment clinically to a sixty-five-year-old man suffering from end-stage coronary disease, complete heart block, and recurrent cardiac arrest. External stimulation was successful and the patient survived for six months.

His work was published in the *New England Journal of Medicine* in 1952. Even though it was praised by the journal's editors, some of Zoll's colleagues thought his work was "against the will of God." *The Pilot,* a Catholic newspaper, intervened, telling its parishioners not to worry about the "outlandish treatment" at Beth Israel Hospital because "God works in many strange ways; this is one way of expressing the Divine Will."

Electrical stimulation of the heart for resuscitation from ventricular standstill was Zoll's fundamental discovery. He introduced external electrical countershock as a basic method of resuscitation from cardiac arrest owing to ventricular fibrillation.

In his 1952 paper on external electrical stimulation, he suggested it should be possible to resuscitate a patient from cardiac arrest by applying externally a strong countershock. In 1956, he developed a clinically practical, safe technique that was successfully applied on humans. Between 1960 and 1964, a method was developed by Zoll and his colleagues for long-term direct electrical stimulation of the heart by an implanted pacemaker. Implanted pacemakers were also employed to improve congestive heart failure in patients with a slow ventricular rate.

By the mid-1950s, cardiac pacemakers with skin electrodes were used to stimulate the heart; these devices, however, left uncomfortable burns that lasted several days. Electrode wires through the skin were tried but infections along the wires were a problem. Ake Senning suggested implanting the entire pacemaker. Senning's suggestion led Rune Elmqvist to design the first implantable pacemaker, which included a pulse generator that delivered about 2 volts with an impulse period of 2 milliseconds. The original transistors showed large leakage currents and were found not suitable. Two newly developed transistors were used in its place. The charging current came from a line-connected, vacuum tube, radio-frequency generator with a frequency of 150 kilohertz. Theoretically, one overnight charging was enough for about four months, but in reality it only lasted for one month.

Arn Larsson, a forty-three-year-old patient who suffered from life-threatening Stokes-Adams seizures requiring him to have thirty resuscitations per day, was the recipient of the first implanted pacemaker. Senning implanted his pacemaker in 1958. Larsson had no complications and was able to lead an active life.

62

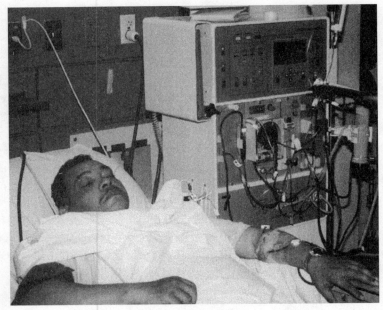

People live years longer because of this device. *National Kidney Foundation*

Kidney Dialysis Machine

Though the kidney dialysis machine cannot replace someone's kidneys, it can give people a means of staying alive and healthy while they wait for a transplant and, at the same time, let them live a quality life.

When a kidney fails, there are two types of treatment available. Most people have what is known as hemodialysis done. This begins when the doctor makes what is called an access into the patient's blood vessels. This may be achieved with minor surgery in the leg, arm, or sometimes neck. A popular method is for a surgeon to join an artery to a vein under the skin to make a larger vessel, forming a graft.

Two needles are then inserted in the resulting graft, on the vein side and on the artery side. Blood then drains into the dialysis machine to be cleaned. The machine has two parts. One is for a fluid called dialysate and the other is for blood. The two parts are separated by a thin, semipermeable membrane. As blood passes on one side of the membrane, and dialysate on the other, particles of waste from the blood pass through microscopic holes in the membrane and are washed away in the dialysate. Blood cells that are too large to pass through the membrane are returned to the blood stream.

Peritoneal dialysis is the other, less frequent treatment, which uses the patient's own peritoneal membrane as a filter. The peritoneum is a membrane that lines the abdominal organs. This membrane—like the dialysis machine membrane—is semipermeable. Waste products can pass through it, but larger blood cells cannot.

First, a plastic tube called a peritoneal catheter is surgically implanted into the belly. About two quarts of dialysate fluid is then passed through the catheter into the abdomen. As the patient's blood gets exposed to the dialysate through the peritoneum, impurities in the blood are drawn through the membrane walls and into the dialysate. After 3 or 4 hours, the dialysate is drained and fresh fluid poured in. The procedure takes about half an hour and must be repeated about five times a day.

The benefits of hemodialysis are that the patient requires no special training and he or she is monitored regularly by someone trained in providing dialysis. The main benefit of peritoneal dialysis is freedom—the patient doesn't have to stay at the dialysis clinic several hours a day, three times a week. The dialysate can be exchanged in any well-lit, clean place, and the process is not painful. The one potential drawback is the risk of infection of the peritoneal lining.

Children often do a similar type of dialysis called continuous cycling peritoneal dialysis. Their treatments can be done at night while they sleep. A machine simply warms and meters diasylate in and out of their abdomens for 10 hours continuously. This leaves them free from treatments during the day.

Although dialysis is still not a cure for kidney disease, and the technology of the dialysis machine remains about the same, great strides have been made to make the dialysis patient more comfortable and mobile.

If a person has chronic, end-stage renal disease, a kidney transplant is the only long-term solution that frees a patient from dialysis. Living relatives can donate a kidney if the remaining organ is healthy. Even with a kidney from a close relative, however, a transplant recipient must take drugs to suppress the immune system from rejecting the organ. At the present time, there are about three times as many people waiting for transplants as there are kidneys available. This translates to many people in need of dialysis as a life-saving procedure.

Some patients, however, simply refuse transplants and see dialysis as something of a social gathering and a way to be monitored and cared for by a group of health care providers that become like friends.

Whether a person chooses to have a transplant or undergo dialysis, the statistics about dialysis and the benefits of the modern dialysis machine are indisputable. Dialysis survival in the United States after one year is 77 percent, according to the National Center for Health Statistics. After five years, it is 28 percent, and after ten years it is about 10 percent. Transplant survival rates are even higher: 77 percent of patients survive ten years after a living-relative donor. Many experts point out there is room for improvement in the survival rate and quality of life for American dialysis patients. More improvements in the machine are expected in the near future.

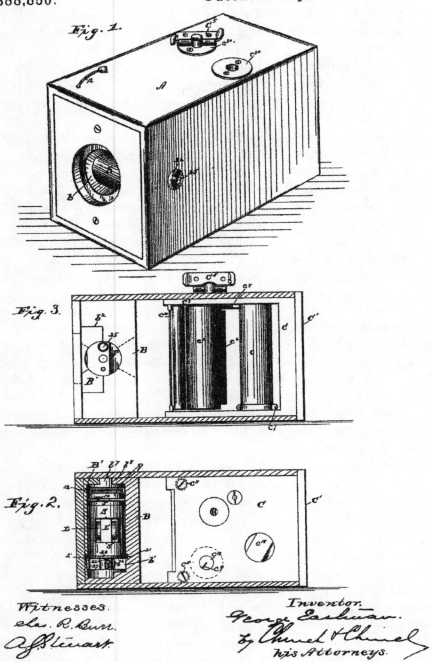

Patent drawing of box camera, 1888, by George Eastman. *U.S. Patent Office*

Camera 63

The story of the camera actually begins with a hunger to record things—scenes, sights, and people—with particular accuracy. Enter the device called the "camera obscura." The term literally means "dark chamber," taking its name from a dark, blackened room or space into which was drilled a tiny hole. Light came through the hole and projected an inverted image onto a light-colored, reflective wall. This was and remains a common way to view solar eclipses without risk to the eyes, but in later times it came to be used by artists who traced the image to create an exact representation of the subject.

Gazing long enough into an image produced by the "camera obscura," one can easily imagine an individual yearning to preserve the projection. One such person was Joseph-Nicéphore Niepce, an amateur lithographer (an artistic painting process).

While not an artist, Niepce was blessed with an inventive mind. In 1822, using a kind of asphalt called bitumen of Judea combined with lavender oil, he exposed a plate to sunlight filtered through a transparent engraving. Where the sunlight touched the bitumen, it became solid and fixed. The darker parts, meaning those not exposed to as much light, simply washed away.

Niepce worked on and perfected his process. He then took his camera obscura and combined it with this process. By attaching a treated pewter plate to the camera, he recorded the image of his courtyard as seen from his house in 1826. It took over 8 hours to expose the image, but when he was done, he had the first photograph ever. Perfecting this procedure, Niepce was able to fix images to lithographic stone, glass, zinc, and pewter and called it a "heliograph" or "sundrawing."

Another person who yearned to fix the image was Louis-Jacques-Mandé Daguerre, a prominent and extremely talented theatrical scene painter who employed the camera obscura extensively in his work. As soon as he heard of Niepce's success with his heliograph, Daguerre sought him out. Daguerre had tried for years without success to fix the image and knew he needed help.

It had already been established by at least 1727 that "silver salts" were capable of responding to sunlight by darkening. Johann Heinrich Schulze in that year demonstrated this by recording words with sunlight, but no one had been able to successfully retain those recordings.

Daguerre tirelessly experimented with this potential but to no avail, that is until Niepce agreed to meet with him. Niepce had taken the heliograph about as far as he could; his dream of producing the effect on paper seemed unattainable since the images were underexposed and far too faint to be etched.

Making a partnership, the two continued their work until Niepce's death in 1833. Carrying on alone, Daguerre used his partner's latest techniques, which included silvered copper plates. By accident, in 1835 Daguerre discovered that an image exposed on a silver iodide plate actually became visible after the fact when introduced to mercury vapor. Unfortunately, the image was fleeting, as the unexposed areas turned dark over time. Still not one to give up, two years later Daguerre was able to remove the unexposed silver iodine from the plate by dipping it in a solution of common table salt, thereby fixing the image permanently.

And what an image it was. Indeed, never before had an image been recorded with such complete detail and exactness! Its inventor gave it his own name: the "daguerreotype."

But this was just the beginning: while Niepce and Daguerre were developing their processes, so were many others. Two notable figures were in Great Britain: Thomas Wedgwood, son of the great ceramicist, and William Henry Fox Talbot.

Wedgwood worked extensively with silver nitrate–soaked pieces of paper and leather but never got very far beyond recording shadows. Talbot, on the other hand, a scientist by trade, gazed through his camera obscura and yearned for a permanent image without knowing about either the French or Wedgwood's efforts.

Then, Talbot actually took photography a step closer to Niepce's dream by creating a light-sensitive paper that had been soaked alternately in silver nitrate and salt solution, making silver chloride. When exposed, the silver chloride created a "negative" image that could then be exposed again and printed in a "positive" image.

What the Talbot process meant was that the same negative could be used to print multiple positive images, unlike each daguerreotype, which was unique and impossible to duplicate. The drawback was that compared to the daguerreotype, the image quality of Talbot's "photogenic drawing" was distinctly inferior, because paper negatives produced inferior images since the details were affected by the fiber in the paper, something glass negatives didn't do.

Regardless of which, it was these two processes that led to modern photography. The daguerreotype also held its own for many years. Gradually, photography spread to everyone, and it was accepted for portraits.

People who traveled loved taking photos as well. Not only did they travel to the U.S. West, for example, but they returned to the East Coast with photographs of the new frontier, strange natives, hard-pressed settlers, and massive vistas of prairies, mountains, and deserts, the likes of which had never been seen before.

There's no better way to describe this journalistic impact than to mention the work of one of its greatest pioneers: Matthew Brady's photographic record of the U.S. Civil War. These images, now so familiar from history books, brought the carnage of distant warfare closer to home than ever before, warfare recorded in unromantic, unglamorous tones.

The "collodion" process was the next major technological advancement on the horizon that reduced the time it took to take a photo even more. It had a quality that rivaled the daguerreotype and was printable on Talbot's paper, which was eventually replaced with "albumen." The only problem was that the plates were made from glass, which needed to be prepared just before exposure and developed immediately after, hence giving it the name "wet-plate" photography.

The image revolution was in full swing in the 1870s. Everyone wanted photographs and it became easier and easier to get them, whether it was for representation and documentation or the exploration of a new artistic medium. Inventors set their sights on all the varied aspects of photography, but the most important of all at this point came in the form of making a "dry-plate" process, wherein the plates could be prepared well in advance and developed long after they had been exposed. The answer came in the form of suspending the silver bromide in gelatin and the result was astounding. Not only was it more convenient to use, but the gelatin proved no less than sixty times faster than the collodion process. A photographer could stand, hold the camera without a tripod, and "snap" a photograph in no time at all.

One pioneer in the use of the dry-plate process was George Eastman. In 1888, the most popular camera around was the Kodak, which was produced by Eastman. He invented the phrase, "You press the button, we do the rest," and he meant it. All the owner of a Kodak camera had to do was shoot all the negatives and mail the camera back to the factory, where technicians developed the pictures. The rest is history as the technology advanced to the point anyone could use a camera and soon anyone could afford one.

64

A portable GPS receiver. *Standard Horizon*

Global Positioning System

Ferdinand Magellan and Christopher Columbus would have had a much easier time and history would have been written very differently if the global positioning system (GPS) was around in the fifteenth century.

The GPS consists of twenty-four Earth-orbiting satellites that allow any person who owns a GPS receiver to determine his or her precise longitude, latitude, and altitude anywhere on the planet. The invention has changed the way people travel—and probably forever.

Vic Beck, a retired navy officer and helicopter pilot, uses a portable GPS receiver on his boat to navigate his way home in foggy weather and the next day takes it in his car to calculate his progress and the distance to his daughter's house among the hills of New Hampshire. As a pilot who was trained with compasses, he enjoys the miracle of the GPS. It is fast becoming a standard feature found in many new cars.

The GPS uses "trilateration," a geometric principle that allows a person to find his or her position from three other, already known locations. An example of how the GPS works could start with a question. Suppose you are totally lost somewhere in the United States. When you ask where you are, another person answers that you are 625 miles from Minneapolis, Minnesota. Though helpful, this information is not useful to you by itself. It simply means that you could be anywhere from Minneapolis within a 625-mile radius. You ask again. This time you receive an answer that you are 690 miles from Boise, Idaho. When you plot these two cities on a map with their surrounding radiuses, you have two circles that intersect. You now know that you are at one of two points, but you don't know which one. You ask again and are told that you are 615 miles from Tucson, Arizona. You can now figure out which of the two points you are at. With three known points, you can see that you are near Denver, Colorado. This is where the three circles on your map intersect.

Although the idea is easy to understand in a two-dimensional space (longitude and latitude), the same concept works in a three-dimensional space (longitude, latitude, and altitude) as well. In three dimensions, the system works with spheres instead of circles. Instead of three circles, four spheres are needed to determine one's exact location.

The core of the GPS receiver is the ability to find the receiver's distance from four (or more) GPS satellites. So, once the receiver determines that, it can calculate its exact location and altitude on Earth. To put that another way, to find your location the GPS receiver has to determine the four satellites above you and calculate the distance between you and each of those satellites. The receiver measures the amount of time it takes for the signal to travel from the satellite to the receiver. Since it is known how fast radio signals travel—at the speed of light (186,000 miles per second)—the receiver can figure out how far they've traveled by determining how long it took for them to arrive.

A very important element of the GPS calculations is the knowledge of where the satellites are. These satellites travel in very high, predictable orbits. The GPS receiver stores information, much like an almanac, that tells it where every satellite should be at any given time. A special feature is that the U.S. Department of Defense constantly monitors the satellites' exact positions and transmits any adjustments to all GPS receivers as part of the satellites' signals. The most important function of the GPS feature is to pick up the transmissions of at least four satellites and combine the information in those transmissions with information in the almanac, so that it can mathematically determine the receiver's position on Earth.

The basic information a receiver provides is the longitude, latitude, and altitude of its current position. The ideal situation is when the receiver combines this data with other information, such as maps stored in the receiver's memory. This is helpful to hikers and travelers of every kind because the whole Earth has been mapped, so there is no place on Earth where you cannot find yourself—and know exactly where you are. Simply put, by applying already known geographic data that has been accumulated to someone's current position or a desired location, he or she can calculate the speed, distance, and estimated arrival time when traveling.

Despite the technology for the GPS having been around for years, the system was originally designed and operated by the U.S. military. In fact, it is funded by the U.S. Department of Defense. It is estimated that there were thousands of worldwide nonmilitary GPS users by 1999, and this number has been increasing since then.

Currently, GPS receivers are used for navigation, positioning, time dissemination, and other research. They are also found in planes, trains, cars, boats, and many other types of vehicles (including hand-held devices).

Sewing Machine 65

The sewing machine may hold the record for the longest time an invention was in development. The first known mechanical sewing machine patent was a British patent issued to a German named Charles Weisenthal in 1755, but archaeologists tell us that people have been hand sewing for twenty thousand years.

The first sewing needles were made of animal bones and the first thread from animal sinew. Later, in the fourteenth century iron needles were used. The eyed needles we know today weren't invented until the fifteenth century.

Although Weisenthal's patent was issued, little is known about his machine. What is known is that it was designed for a needle that was meant to be used in mechanical sewing. The same fate befell English inventor and cabinetmaker Thomas Saint. Although he was issued a patent for a sewing machine in 1790, it is not known if Saint developed a prototype machine. The patent describes an "awl" that punched a hole in leather and passed a needle through the hole. Later, because there was a description of the machine but no evidence that one existed, other inventors set out to build his machine. The reproduction, based on his drawings, did not work.

The next adventure in sewing occurred in 1804 when a French patent for a sewing machine was granted to Thomas Stone and James Henderson that was said to "copy" hand sewing. This invention failed not long after and was forgotten.

In 1818, the first American sewing machine was invented by John Doge and John Knowles. The problem with their machine was that it malfunctioned so often, even after trying different kinds of fabric, that it was more of a "fix-it" machine than a reliable sewing machine.

Not long after, in 1830, the first functional sewing machine was invented by a French tailor. The problem he had, however, was considerably worse than mechanical. His problem was staying alive long enough to enjoy the success of his invention. He was nearly killed by fellow tailors who burned down his sewing machine factory because they feared his invention would put them all out of work.

This same fear—that his invention would cause unemployment because it was able to do the work of many tailors—influenced an American inventor named Walter Hunt in 1834. Too bad. His machine could sew straight seams and was very reliable.

Finally, the sewing machine's fate turned more favorable. The first American

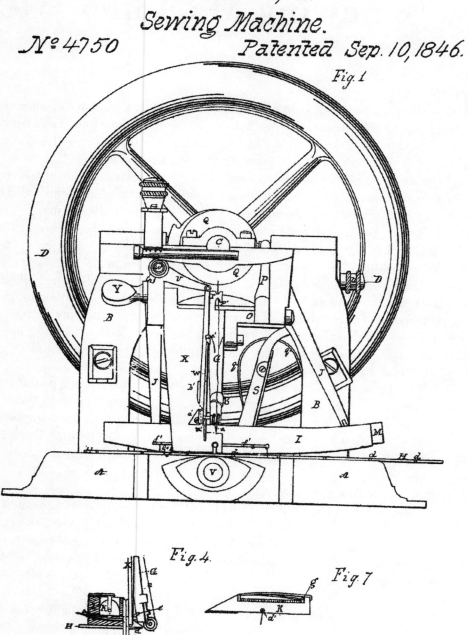

Patent drawing, 1846, by Elias Howe. *U.S. Patent Office*

patent for one was issued in 1846 to Elias Howe. His unique patented process did something no other machine had done in the past: It had a needle, with an eye at the point, that used thread from two different sources. The needle was pushed through the cloth and created a loop on the other side. A shuttle on a track then slipped the second thread through the loop. This process created what became known as a "lockstitch."

Although his machine was useful and practical, his invention wasn't meant to be. Howe struggled for the next nine years, first to generate interest in it and then to protect his lockstitch mechanism from imitators. But he couldn't. His method was taken and used by others who improved it. From Howe's original mechanism, an up-and-down mechanism was developed by Isaac Singer and Allen Wilson developed a rotary hook shuttle.

The style developed by Singer led to the first commercially successful sewing machine. For what Howe felt was theft of his idea, he sued Singer because he maintained that Singer's machine used the same lockstitch that he had invented.

Meanwhile, Singer had added other invaluable features, such as operating the machine by a foot pedal rather than by an earlier hand-cranked version. Its obvious advantage was keeping the operator's hands free to work. Another improvement was that the needle moved up and down, entering the fabric being sewed from the bottom and the top.

Although Howe won his lawsuit for patent infringement against Singer in 1854, sewing machines did not go into mass production until much later. The first mechanical sewing machines were used in garment factory production lines, and it wasn't until 1889 that a sewing machine for use in the home was designed and mass-produced. By 1905, an electrically powered sewing machine was in wide use.

Meanwhile, after successfully defending his right to share in the profits of his invention, Howe saw his wealth jump to more than $200,000 a year soon afterward. From 1854 to 1867, Howe earned close to $2 million from his invention. During the Civil War, he donated a large sum of money to equip an infantry regiment for the Union army and served in the regiment as a private.

The influence of the sewing machine cannot be overstated. Overnight, it changed the clothing and other industries where sewing was required and consequently changed the lives of millions of people who were ultimately part of that industry.

66

Rolls of photographic film. *Author*

Film

Film is one of those inventions whose true importance to humanity gets somewhat lost in the shuffle. It tends to be taken for granted, just something that exists. When one contemplates the various things it does (with the help of a camera, of course) it's quite impressive.

Consider, for example, the job law enforcement personnel would have at crime scenes without being able to take photos. Or consider its military and scientific uses, where it's used to preserve facts in an instant that we would have no other practical way of recording. Also consider its role in the creation of books and magazines. It's simply unimaginable to think of them without pictures. And, of course, what about our memories? Think about how much pleasure and joy we derive in viewing pictures of friends, family, and pets of years past.

Film works as it does because it uses chemicals sensitive to light, which is the visible portion of a broad range of electromagnetic radiation that actually includes invisible energy in the form of radio waves, gamma and X rays, and infrared and ultraviolet radiation. The relatively narrow band of electromagnetic waves the human eye can detect is called the visible spectrum, which we know as color. The human eye perceives the longest visible wave lengths as red, the shortest as violet, with orange, yellow, green, and blue in between. Prisms and rainbows display all the colors of the visible spectrum.

The photosensitivity of certain silver compounds, particularly silver nitrate and silver chloride, was recognized by the 1700s. In England in the early 1800's, Thomas Wedgwood and Sir Humphry Davy tried using silver nitrate to transfer a painted image onto leather or paper. They managed to produce an image, but not a permanent one; the surface blackened when it was exposed to continuous light.

In France, Joseph-Nicéphore Niepce made the first successful photograph in 1826 by placing a pewter plate, coated with bitumen (a light-sensitive material), in the back of a "camera obscura." Niepce later used copper plates and silver chloride in place of bitumen and pewter. In 1839, after Niepce's death, Louis-Jacques-Mandé Daguerre demonstrated an improved version of the process that he called a "daguerreotype."

Daguerreotypes produced images on a shiny, hand-held copper plate. Though popular, they were replaced in the late 1850s by a negative-positive process devised in England by William Henry Fox Talbot. Talbot exposed silver-sensitized paper briefly to light and then treated it with other chemicals to produce a visible image; from this negative, a number of positive images could be produced. Starting in 1850, glass replaced paper as a support for the negative. Silver salts were suspended in "collodion," a thick liquid. Glass negatives produced sharper images than paper ones because details weren't lost in the texture of the paper. (This became known as the "wet collodion" or "wet plate" process.)

Because the wet-plate process meant the glass support had to be coated just before a picture was taken and developed immediately after, a "dry-plate" version of the same process was sought. Dry plates, pieces of glass coated in advance with an emulsion of gelatin and silver bromide, were invented in 1878. After that, American George Eastman devised a flexible system: a long paper strip replaced the glass plate. In 1889, Eastman used a plastic called celluloid instead of paper; this was the first photographic film. Eastman paved the way for all films that are now made of polyester or acetate, plastics that are not as flammable as celluloid.

Except for a few isolated experiments, color films were not used until the twentieth century. Autochrome, a commercially successful material used for making color photographs, became available in 1907, based on a process devised in France by the inventors Auguste and Louis Lumière, but the era of color photography only really began with the advent of Kodachrome color film in 1935 and Agfacolor in 1936. Both produced positive color transparencies, or "slides." Eastman-Kodak introduced Kodacolor film for color negatives in 1942. This gave amateurs access to a readily available negative-positive color process.

Photographic film uses chemicals that vary in the way they react to different wavelengths of visible light. Early black-and-white films used chemicals sensitive to the visible spectrum's shorter wavelengths, primarily light perceived as blue. In an early picture of colored flowers, blue ones appeared light, while red and orange ones looked too dark. To correct this, specialized compounds called dye sensitizers were incorporated into the emulsion, thus recording colors as different shades of gray. Today, with a few specialized exceptions, films are sensitive to all colors of the visible spectrum.

67

When this was invented, people tried to kill its inventor. *Drawing by Lilith Jones*

The Spinning Jenny

As legend would have it, when little Jenny Hargreaves knocked over the family spinning wheel in the mid-1700s, two very important things happened. First, James Hargreaves, her father, gazing into the still revolving spindle, experienced one of those intensely dramatic moments of inspiration of which most inventors only dream. Second, he translated this inspiration into reality by creating the first workable improvement on a machine and system that had been in use by people since the dawn of civilization: the process of making "thread."

In the beginning, individual fibers were pulled from a clump of wool placed on a forked stick called a "distaff" and wound onto another stick called a "spindle." The thread created by this process could then be woven on a "loom" to make fabric.

Over the years, improvements were made. One included mounting the spindle so it turned on bearings, driven by a cord attached to a wheel spun by hand, thus eliminating the need for a two-person operation. This device came to be known as the "spinning wheel." Over the centuries, the spinning wheel spread readily throughout the world, with the only substantial design improvements being a foot pedal and a mounted distaff allowing for hands-free operation. This particular machine was called the "Saxon wheel" or "Saxony wheel" and first appeared in Europe in the early 1500s. While the Saxon's design allowed a greater output of good-quality wool or cotton thread and yarn, it still took between three and five wheels to keep one loom going. That is until John Kay.

Kay had the distinction of inventing the "flying shuttle," an important innovation in the development of the loom. In short and simple terms, the loom operates by holding two sheets of yarn or thread taut. This is called the "warp." The sheets of warp are held apart by the loom while a "shuttle" passes between them cross-wise, unwinding a length of thread as it goes. This thread is called the "weft." From the time the ancient arts of spinning and weaving were conceived, the shuttle or its operational equivalent were passed by hand through the sheets of warp until Kay's flying shuttle appeared, which mechanically hurled the shuttled weft through the warp twice as far and much, much faster. This improvement produced fabric twice as wide and far more quickly than ever before, creating, understandably, a sudden and profound shortage of thread and yarn.

Now James Hargreaves was a simple man, living at the dawn of the Industrial Revolution. He and his family were part of the "cottage industry," meaning that the whole family supported itself by the weaver's trade. From its home, the Hargreaves family processed fiber into thread, then thread into cloth. An enterprising cloth merchant sold the Hargreaves the raw material—wool fleece or cotton—and returned sometime later to purchase the finished cloth, while selling more raw material to the family.

Hargreaves built a machine that one person could operate to spin several threads at once. It began with eight spindles in 1764, with a potential seen for many more, and was an instant success as far as the merchants and entrepreneurs were concerned. Hargreaves began to supplement his family's meager earnings by making and selling the machines.

When technological innovations like the flying shuttle or the spinning jenny came along that increased production, required less labor, or both, as in this case, the results to the simple cottage workers could prove devastating. For this reason, the Hargreaves's spinning neighbors had little reason to rejoice and much to fear. In 1768, their home was attacked and all the machines destroyed.

Undaunted, the family moved to Nottingham to found a modest spinning mill based on his machine. But, in the fast-paced, dog-eat-dog era of invention and innovation, James Hargreaves, the humble weaver from the village of Stanhill, did not receive the success he really deserved.

He had every reason in the world to be immensely proud of the first major advancement for centuries in the art of spinning, but between 1764 and 1770 the courts refused to give him a patent until he had built the sixteen-spindle version. Unfortunately, Hargreaves had made and sold too many of the eight-spindled kind, that by the time he got around to filing for a patent he was flatly refused because a multitude of copies on his design had come into being. It's estimated that at the time of his death in 1778, over 20,000 spinning jenny machines were in use in Great Britain alone, with as many as 120 revolving spindles.

However, the story of the spinning jenny and the textile manufacturing boon of the 1770s was hardly over at that point. For the crafts of spinning and weaving, the spinning jenny and the flying shuttle were just the beginning. Further advancements on spinning and weaving loomed on the horizon, less than a decade away, advancements so profound that they changed society and the way it used labor forever.

Sir Richard Arkwright started out as a wig maker but eventually set himself to the tasks that lay ahead in textile production. Coming from a completely different background than Hargreaves, Arkwright saw something different in the spin of the machine. What he saw and patented in 1769 was the "water frame," a textile machine powered not by hand, but by water. By employing teams of engineers, including John Kay, Arkwright set his sights on improving and mechanizing the entire industry from carding to the loom.

In seeking to enhance production and reduce labor costs, Arkwright brought his huge machinery to large buildings near water sources where waterwheels moved the machines. This resulted in the need for workers, mostly women, to come to the spinning apparatus instead of working at home. This effort resulted in the first textile "factories" or "mills" (because of the water propulsion), later to become known as "sweatshops."

What Arkwright and others who came after, like Samuel Crompton, also accomplished was to correct a major flaw with Hargreaves' spinning jenny, which unfortunately made cotton and flax thread that was easily broken. This thread, especially if it was wool, could be used for the weft on the loom, but it was too fragile for the warp. Arkwright's water frame had no problem, however, and lasted until Crompton, himself an uneducated mill worker, built a better spinning jenny in 1779: the "spinning mule." Crompton never patented the mule and only received a small amount from manufacturers when he sold the design to them. In March 1792, the same year Eli Whitney invented his "cotton gin," a group of angry spinners broke into the Grimshaw factory in Manchester and destroyed all the spinning mules installed there.

While the spinning jenny may almost seem obscured in the flurry of textile inventions of the eighteenth century, it stands out as an invention that began the ushering out of one era and the beginning of a new one. Regardless, wherever the cottage industry still flourished, straight through to the middle of the nineteenth century, one could still find the spinning jenny spinning away.

The brick: convenience, strength, beauty. *Author*

Brick

Behold the brick: Not much in and of itself when you look at it. Ordinarily rectangular in shape, it has ranged in size anywhere from the convenient hand-held variety (very popular in street riots) to the enormous, pyramid-building kind thought to have been moved by hundreds of slaves at a time. It just is not much to look at or think about and is often used as a metaphor to describe particularly dense people who don't listen well. But when put together in groups and placed into structures throughout history, especially in the hands of gifted masons, the brick is one of the most fundamentally gorgeous inventions of all time.

The story of the brick is believed to have begun in the birthplace of civilization, along the banks of the Tigris and Euphrates rivers, although this scene could have just as easily played itself out in China, Africa, Europe, or wherever early humans were encamped. As floodwaters receded, deposits of thick silt and sediment were left behind and exposed to the sun. As this "clay" dried, it caked and cracked into smaller pieces that could be shaped into any number of desired forms, like small statues, bowls, and bricks (which were subsequently stacked to make rudimentary huts, sheltering humanity from the very elements that created the bricks). It amazes one to think that these simple, rough-hewn items—sculptures, housewares, and bricks, all gifts from the earth, rivers, and the sun's rays—all share the same technological evolution.

There is a very early written account that describes the first true "arch" made of these bricks dating back to 4000 B.C. in the ancient Mesopotamian city of Ur, in what is now modern Iraq. The arch itself is a critical element to the evolution of building and architecture, made possible by the brick. By shaping a brick into a "wedge" and putting several of these wedges together, the mason was able to distribute the weight of a structure evenly between each brick. This combination of human ingenuity, physics, and sheer strength of the brick literally gave rise to everything from the simple doorway, to bridges and aqueducts, to the Roman Coliseum, to vaulting Gothic cathedrals. It replaced for the most part the massive, relatively unstable "lintel," made of an unwieldy, horizontal stone slab that was balanced precariously on top of two vertical slabs.

While that first Mesopotamian arch has long since returned to dust, the ancient description of it includes the first known reference to what has become over time one of the most important modern building materials. Those bricks were held together with a thick, tarry mortar called "bitumen slime," a great-great forebearer of our modern asphalt.

By containing fires used for cooking and warmth, early hearths provided the next major development in the brick. Clay bricks, when exposed to the heat of the flame, took on a much stronger stonelike consistency. In the Mesopotamian city of Ur, we find that the sun-dried bricks of that great city were eventually replaced by "fired" ceramic ones. The potters of this ancient culture developed an oven capable of producing a very high heat, called a "kiln," in which to bake their wares. The kiln provided the potters with a more even, controlled method of baking or firing the mud into a hard, heat-resistant ceramic. The temperatures to achieve this ranged anywhere between 1,600 and 2,000 degrees F or more. Thus, the brick in the more modern sense came into being around 1500 B.C.

In turn, keeping the close relationship with the potter's and sculptor's craft, bricks were the material from which kilns were and are still made. As the brick evolved throughout history, so did pottery and clay sculpture methods. The next evolutionary step for the brick, in fact, the "tile," employed advanced ceramic techniques such as "glazing," which provides the surface with not only a wide variety of colors, but also a smooth or even shiny texture that is nonporous. There remains to this day no material that matches brick and tile for endurance, strength, and resilience to even the harshest of acids. As demonstrated by the equally durable story of the "Three Little Pigs," when using a shelter made of brick, little else compares.

As the art of brick making reached out of the cradle of civilization and throughout the world, material, style, and technique became as unique as the people and cultures that took up the craft. From the earth were derived a variety of different clays that ultimately found their way into the colorful character of each culture as well as providing varied characteristics to the architecture. The three basic types of clay are: (1) surface clay, which is usually found in river bottoms and used fairly easily; (2) shale, which is highly pressurized rock, like slate, found in various stages of firmness; and finally (3) fireclay, which is mined from deep underground and is more pure in composition than the other two.

Throughout the Mediterranean world, bricks found their zenith in the form of "terra-cotta," meaning literally "baked earth." Bricks were used extensively in Roman construction with their cousin "concrete" to create such marvels as the Pantheon in A.D. 123 with a previously unheard of brick and concrete dome that measured 142 feet high.

As Europe emerged from the Dark Ages and cities grew dense with people and wooden structures to house them (it's hard to remember at this point in history how thickly forested much of the European continent was), fire could prove devastating, like the Great Fire of London in 1666. The brick's natural flame resistance made it the desired material from which every great city was built.

Moving east, we find structures as immense as the Great Wall of China, the monumental defensive structure whose construction lasted from the seventh through the fourth century B.C. By the third century B.C., Shih huang-ti, the first Chinese emperor, connected a loose group of walls into one massive fortification that stretched a mind-boggling 4,160 miles to protect the people of his land from invaders, thus allowing the various provinces behind the wall to unite into one China. Interestingly, the Great Wall, visible from space, is actually a powerfully crafted combination of fired and sun-dried bricks.

While this brings the story of the brick a little further away from being the central structural component it started out as, it highlights the fact that, even in its simplest form, the basic red, 2¼-by-3¾-by-8-inch brick still makes the loveliest and most attractive construction material and a superb metaphor for the endurance of humanity.

Motion Picture Camera   69

The impact motion pictures have had on America and the world is impossible to measure. Movies are such an integral part of American culture that fads and fashions among young people are often dictated by them, and they have also exerted wide and deep influence on our likes, dislikes, and belief systems.

Motion pictures are really nothing more than photographs flashed in front of our eyes so quickly we cannot distinguish when one is taken away and the next is put in its place. Thus, these pictures seem to be in motion, in real time.

This phenomenon is called "persistence of vision" (an image remains in our vision for a millisecond) and has been well known for quite some time. But only around the 1880s did people start making a machine that could do this and let us view our own reality.

Although somewhat crude animation and lantern slide shows existed early on, the goal of these inventors was to capture reality much like it is seen by the participants. The first known experiments with motion pictures occurred when Eadweard Muybridge used a series of cameras to record animal and human locomotion. To do this, he created the "zoopraxiscope" or what was also called the "wheel of life."

The machine worked by moving drawings or photographs that were viewed through a slit in the zoopraxiscope. Although it wasn't really what we consider motion pictures today, it was the state of the art at the time. Muybridge is considered by some historians as the Father of the Motion Picture. What really put Muybridge on the map was his most famous short film, where he used a row of cameras that snapped a dozen pictures of a passing horse. The public was astonished to see proof that a trotting horse can simultaneously have all four hooves off the ground.

Later, Muybridge devised a fast camera shutter and used a new, more sensitive photographic process. The result was that it dramatically reduced exposure time and for the first time produced consistent, crisp images of moving objects.

Some disagree with Muybridge being credited with inaugurating motion pictures because there were several people working on devices for motion pictures at the same time, including Thomas Alva Edison. He and W. K. Laurie Dickson, his assistant, originated one of the first practical systems in the late 1880s. It

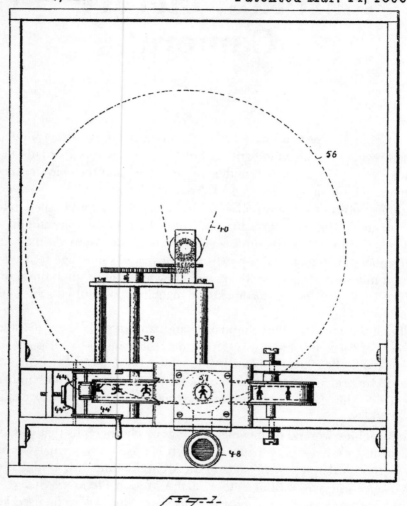

Patent drawing of motion picture device, 1893, by Thomas Alva Edison. *U.S. Patent Office*

used a motion picture camera that he called a "kinetograph" and a viewing system he called a "kinetoscope." Both devices were patented in 1891.

Edison and Dickson didn't stop working there. The world's first movie studio, the Black Maria, was built under Dickson's direction between 1891 and 1892 and many short films were made there. One of these included the first "Western" movie *Cripple Creek Bar-Room* in 1899.

The movie studio's success exploded and interest in seeing these short films soared. By 1892, these short movies could be seen at penny arcades, peep shows, or kinetoscope parlors.

Eventually, projectors were made that enabled a theatre owner to present films to a much larger audience. The public's interest and curiosity were insatiable. Movie theatres sprang up all over the country and the world. Audiences watched film vignettes and short films and wanted more. Many of the films were of city scenes and nautical environments, while a popular one was of a train coming right at the camera.

Again, another step forward in the business of motion pictures has Edison's stamp on it. Foremost among American filmmakers was Edwin Porter, who shot a number of films for Edison. His *Life of the American Fireman* in 1903 told a real story and later his *Dream of a Rarebit Fiend* in 1906 was both unusual and funny. These films certainly showed other would-be moviemakers of the time the potential of film.

The Great Train Robbery in 1903, the most celebrated early Western film, was a huge hit in its day. It was filmed in New Jersey (the actors weren't really from the West, either). The great part of this movie that viewers adored was its pace. The story involved a daring train robbery that led to the downfall of the robbers at the hands of a posse. The movie simulated a fast train with sound effects and exhaust billowing frantically from the engine car. Moviegoers had seen nothing like it before and couldn't get enough of it.

Another name synonymous with motion pictures is the Frenchman Louis Lumière. He invented a portable motion-picture camera, film processing unit, and a projector called the "cinematograph." It covered those three functions in one machine.

The cinematograph made motion pictures even more popular and some believe Lumière is truly the Father of the Motion Picture. In fact, Lumière and his brother were the first to present projected, moving, photographic pictures to a paying audience of more than one person. I wonder what he would say about what today's audiences are willing to pay.

70

Alfred Nobel, inventor of dynamite. *New York Public Library Picture Collection*

Dynamite

Alfred Nobel, the Swedish inventor, invented dynamite in 1866. His invention, like many others before it, was the result of the accumulated knowledge of different people. His genius comes from his tireless effort and ultimate ability to perfect dynamite's use for predictable, practical applications worldwide.

The story of dynamite starts with Ascanio Sobrero, an Italian chemist, in 1846. He is the first person known to treat glycerol with nitric and sulphuric acid to produce the primary ingredient in dynamite: nitroglycerine. The problem was that by adding nitric and sulphuric acid, heat was created and the mixture became unstable, often to the point of exploding. It took some time before inventors realized the mixture needed to be cooled while being made. This increased its predictability and stability.

Nobel studied these problems and was the first person to manufacture nitroglycerine on an industrial scale. One of his most important discoveries was the result of mixing the now stable nitroglycerine with an oily fluid and silica. The whole mixture could be turned into paste, kneaded like dough, and shaped into rods.

The problem that remained, however, was how to explode the rods of dynamite. Nobel developed the "blasting cap" (a wooden plug filled with gunpowder that could be detonated by lighting a fuse) in 1865, which could be used to detonate dynamite under controlled conditions.

Controlled conditions meant that predrilled holes in bedrock could be filled with the sticks of dynamite and exploded. The resulting exploding and clearing of rock saved hundreds of man hours. This allowed construction to move much more quickly: what many men could do in several days, a stick of dynamite could do in just a few minutes. You can still see parallel lines in rock formations along today's highways, especially in mountainous areas, where holes were drilled for the dynamite.

Since there were many uses for the product and the building of roads and dams were made a great deal faster and easier, Nobel made a king's ransom from the material's use worldwide. The same driving curiosity that motivated him to invent also made him reticent about the future of his invention.

But Nobel, troubled by potentially violent uses of dynamite, decided to leave his fortune to reward people who pursue peaceful purposes. When he died, he left $9 million to establish a series of prizes in his name: the Nobel prizes for medicine or physiology, physics, chemistry, economics, literature, and peace. Efforts to promote peace were important to Nobel and he derived intellectual pleasure from literature and science, which were the foundation of many of his activities as an inventor.

Nobel understood his place among inventors. He saw the future and worked on the accomplishments of those in the past. He also studied the long and complex history of dynamite.

Black powder, the first chemical explosive, was invented in China in A.D. 900. Made from charcoal, sulfur, and potassium nitrate, it was originally used for military purposes. Later on, it was used in mining in Europe. Fire or intense heat was used to detonate the black powder. Eventually, fuses made of grass or vines were used to accomplish this.

In the modern era, nitroglycerin in dynamite succeeded black powder as the chief explosive. Two important modern developments were "safety fuses" and blasting caps. For the first time, these elements allowed safe and accurately timed detonations.

Not being one to rest on his accomplishments, Nobel went on to improve his invention. In 1875, he created a jelly from the dissolution of nitroglycerin. Testing proved that this material not only created a more powerful explosion, but also was safer to use.

The addition of ammonium nitrate—which is normally used as a fertilizer—to the mix made the material safer and cheaper to make. As a result, its use worldwide increased greatly.

The blasting cap provided the first safe and dependable method of detonating nitroglycerin. This not only made it more safe to work with for construction crews, excavators, and builders, but it also opened the door to a variety of industrial uses for the product.

Another development made over time was electrical firing. First used successfully in the late nineteenth century, electrical firing allowed greater control over timing. This had significant effects on safety and practicality. The removal of bedrock, for instance, could be timed with machinery following the explosions for debris removal.

Naval cannon, nineteenth century. *New York Public Library Picture Collection*

Cannon

Cannons are very old weapons technology, dating back before the 1400s. They weren't too complicated, just a strong metal tube with a plug at one end, where a small hole was drilled for a fuse. Gunpowder was pushed down the tube from the open end and a cannon ball followed so it and the powder were packed together at the plugged end. Either a fuse or the gunpowder itself was used to fire the cannon. An explosion shot the cannonball that smashed through walls, ships, or anything else in its way.

The word "cannon" comes from the Latin *canna* (reed) for the cylindrical bore or barrel. Cannons were invented after gunpowder was brought to Europe from China. They remained until the mid-1800s, when they were replaced by breech-loading guns.

The history of cannons starts with gunpowder, which contains several substances not occurring naturally in pure form, such as potassium nitrate (saltpeter). Chinese alchemists in the Sung dynasty (A.D. 900) came across a white, crystalline powder that cooled water when dissolved in it and exploded when thrown on fire. The Chinese only used gunpowder for incendiaries, such as in fire lances, signaling devices, and firecrackers. Trade routes brought this "Chinese snow" to Europe, where various combinations of natural petroleum, sulphur, or other combustibles had been tried for siege warfare with catapults and smoke.

The cannon's ancestor was the fire lance, a several-foot-long bamboo tube drilled through the joints, wrapped with strong twine, and attached to a long heft to hold and aim it. When the fire lance was lit from a fuse sticking out its muzzle, it discharged its fire, gases, and projectiles to the front, much like Roman candles do for us today. Weapons like it were used in China by the 1200s, then spread throughout the Middle East. Turks, Arabs, and Europeans probably developed the cannon from the fire lance. The phrase "Poudre de canon" is recorded in 1338. Alfonso XI used cannons against the Moors in Spain during the 1340s. Venitians used them during a siege in 1380. Turks used them to besiege Constantinople in 1453.

Modern black gunpowder contains potassium nitrate, sulphur, and carbon in the form of wood charcoal (approximately 4:1:1 for strong cannon powder, and 10:1:2 for pistol powder, while an average modern ratio is 75:11:14). The potassium nitrate has to be purified by recrystallization, and the sulphur through distillation. The charcoal powder has to come from perfect, uniform wood. All the ingredients are coarsely ground, put into an iron pot, and moistened with water, alcohol, vinegar, or urine (to extinguish any sparks). The entire mixture has to be constantly bruised by an iron rod (lift and let fall repeatedly for 24 hours in a stamp mill); if it is not mixed this way, it'll burn or fizzle instead of detonating when lit. Powder used to be ground on stone tablets until mills were used as rotating wheels—one water-driven powder mill could make as much powder in the same amount of time as a hundred men with mortars and pestles.

When the technology of fire lances hit Europe with its iron forges and bell foundries, metal replaced bamboo, leather, wood, stone, and any other sort of material used in cannon making. The first metal cannons were made from iron bars welded together side by side and reinforced by iron hoops. They fired stone cannonballs. Later on, they were cast in bronze of gun metal (copper and tin, which is different from bell metal) and bored for stone or lead balls to fit closely without packing.

These cast guns could be elevated. They had better range than the earlier crude cannons but usually were used at a slight elevation, called "point blank." By the end of the sixteenth century, cast iron (cheaper than bronze) had superseded stone for both balls and cannons. If the casting was poor, the cannon could burst the first time it was fired. After casting, the cannon was held in place and rotated while a gearing set a boring tool into an end to make the barrel completely cylindrical. Then a drill was used to make a "touch hole."

The cannon was tapered from base to mouth with rings around the barrel—a decorative feature reminiscent of older hooped barrels. The base of the cannon was provided with a small ball called the cascabel to help point the weapon. To fire a cannon, a measure of gunpowder was shoveled in from a powder cask and emptied at the base of the bore. The powder was rammed firmly behind a wad or plug of wood or cloth that served as both a gasket to contain the gases when the charge was detonated and a piston to push the ball out of the muzzle. After the ball was loaded into the bore and rammed up firmly against the wad, a gunner poured gunpowder into the touch hole. When the command to fire was given, the gunner used a slow match kept burning in a portfire or a red-hot rod to ignite the gunpowder trail from the touch hole. This trail quickly burned down to the charge and ignited it. The charge detonated and discharged the ball, practically at the speed of sound. Before the cannon was reloaded, the bore had to be moistened to extinguish any embers left from the previous firing with a wet swab.

Cannons changed the composition of standing armies, as artillery divisions were added to the already existing cavalry and infantry divisions. The trench mortar, a descendant of the smoothbore cannon, was an infantry weapon used as late as World War I. It was an effective mass-shock weapon, although its placement was critical: a fixed battery was located in a field as open and level as possible and could be quickly directed to fire at any point within range, but it had to be protected from assault. A battery could usually protect itself against a frontal assault by firing directly into the enemy, but it was vulnerable to a cavalry attack from its rear and flanks.

A volley across a cavalry or infantry charge could produce horrible destruction, no matter whether solid shot or a bag of grapeshot, or two half-shells joined by a length of chain were used as ammunition. A cannonball heated red hot could be used to start fires; cannons could reduce castles and fortresses to crumbled ruins.

Cannons quickly became indispensable for naval warfare. Boarding the enemy's ship became the last stage of a battle, not its start. By first maneuvering to gain advantage and bombarding the enemy ship to destroy its masts and rigging so that it was crippled, the attacking ship then had its cannons sweep defenders off the enemy ship's weather decks, stave in its wooden sides, and set it on fire by heating the shot red hot in a furnace and then firing it so that it was lodged in the enemy's flanks. Naval warfare with cannon was not fun for sailors.

Today, of course, cannons are still an important part of warfare, whether placed in a tank, on a ship, or on the ground.

72

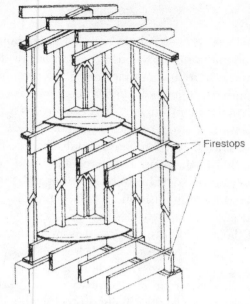

A section of balloon framing. *U.S. Government*

Balloon Framing

Balloon framing—the name refers to its light, airy nature—coupled with the invention of the machine-cut nail, where nails could be mass-produced by machine, revolutionized building in America when it was introduced in the 1830s. Until balloon framing, builders used the laborious "timber" framing, which involved using relatively exotic joinery, such as mortise and tenon, and huge wooden members that could range in size from 4 by 4 to 9 by 15 inches and required many workers to manipulate them into position for erection. It was always a labor-intensive experience and sometimes a backbreaking one.

Balloon framing members ranged in size from 2 by 4 to 2 by 12 inches and were made in various heights (some fairly high) but could be easily handled by one or two people, in that the member was sawed down to the proper size, placed into position, and then nailed. The bottom line was that it allowed buildings to be erected at much greater speed. The *Architectural Review* magazine stated in 1945, "The great cities of the world would not have arisen as quickly as they did if it were not for the invention of the balloon frame, which substituted a simple construction of nails and plates for the old craft of mortised-and-tenoned joints in wooden house construction."

Indeed, the suburbs as we know them could not exist because the houses would take too long to build, and the resulting buildings would be too costly for most people. Moreover, a massive labor force would be required to put the members in place.

Style, of course, would also be limited by the material at hand. For example, if you wanted something with many angles, or perhaps was circular, you couldn't do it without a large income to cover the construction costs.

Also, the sizes of the material available could be a problem. There were only so many tall, thick trees in any given forest. In line with that, the use of trees in such a fashion might have been even more roundly disapproved of by environmental groups then they are today.

The heart of balloon framing is that the "studs," the vertical 2-by-4s that serve to frame the sides of the house, are long enough to extend from the basement to the roof. In general, the building starts with 2-by-6s called "sills," being bolted to the tops of the masonry foundation wall. Then "posts," which are usually made with 2-by-4s, are secured at the corners with the posts made so that there is a recess for accommodating interior wall materials. "Joists," which are composed of 2-inch boards 6 or 14 inches deep, are put in place. Once the joists are installed, the house-high studs are raised into position and made straight (plumbed) and spiked to the sill and adjacent joists.

"Ribbands," also known as "ribbons," are "let"—that is, installed in the slots cut into the studs—and secured to the studs at the second-floor level on opposite sides of the house. These boards are either 1 by 6 or 1 by 8 inches and are used to support the joists, which are laid across the top of and secured to the studs. A pair of ribbands is also secured to the studs at the ceiling/attic height of the second-floor joists. "Plates" are then nailed to the tops of the studs, with each plate consisting of 2-by-4s that have been spiked to the studs and then each other. The plates are cut to fit where they contact the corner joists.

Balloon framing is not as strong as "platform" or "Western" framing, which was developed after balloon framing, so to strengthen it, sheathing—wide, thin boards—are secured to the framework to give it solidity and wind resistance.

One other key element in balloon framing is the "fire-stop," which is a horizontal board nailed between studs at certain points. In a two-story home, there should be two: one at the ribband at the sill and the other at the second-floor ribband. Without these fire-stops, the spaces between studs could act as flues, and a fire would spread much more rapidly.

As inventive as it was, balloon framing on the scale it achieved would not have been possible without the simultaneous emergence of machine-cut nails. For a long time, the only nails available were hand-wrought ones. To make a nail, a blacksmith or nailor had to heat a long thin metal rod called a "nail rod," hammer one end to a point, cut the rod to make it nail sized, and pound the nonpointed end into a head. Making nails was so labor intensive that the cost of building a house could be half of the total materials cost. Indeed, hand-wrought nails were so difficult to make and valuable that the nails were often salvaged when old structures were taken down.

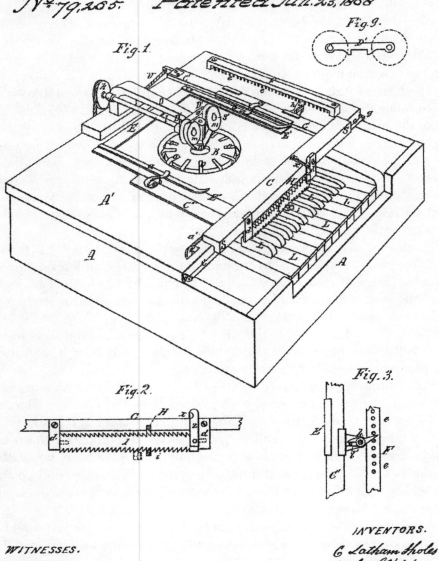

Patent drawing, 1868, by Christopher Latham Sholes, Carlos Glidden, and Samuel W. Soulé. *U.S. Patent Office*

Typewriter 73

It's been said that an army of professional typists working in a room can sound like rapid machine-gun fire. Perhaps appropriately, the first typewriter, called the Sholes and Glidden Type Writer, was produced by the gun makers E. Remington and Sons in Ilion, New York, from 1874 to 1878.

The original model didn't sell that well (fewer than five thousand units), but it accomplished much more: It introduced the machine age into the office, relieving workers of dreary time-consuming work, and started a worldwide industry.

The typewriter appeared in an age when inventors were engulfing the U.S. Patent Office with plans for devices to make life easier for the general public, which was fascinated by machines, especially those that made everyday tasks less tedious. Everyone also seemed to simply be in love with ideas.

The typewriter began at Kleinsteuber's Machine Shop in Milwaukee, Wisconsin, in 1868. A local publisher-politician-philosopher named Christopher Latham Sholes spent hours at Kleinsteuber's with fellow tinkerers, eager to invent something useful.

The story goes that Sholes was working on a machine to automatically number the pages in books when someone in the shop suggested the idea might be extended to a device to print the entire alphabet. Later, an article from *Scientific American* was passed around, and the men agreed that "typewriting" (the phrase used in *Scientific American*) was the start of a growing trend and would have a bright future.

The idea of "extending" ideas from existing machines to new ones not yet built was an important characteristic of the time. It required inventors to sometimes make ordinary objects extraordinary by combining them with others for some important purpose.

Sholes, of course, accomplished just this with the typewriter. His earliest demonstration model was a simple device that used the old key of a telegraph instrument mounted on a base. It was a kind of "let me show you what I hear" jump from the simple telegraph idea.

Sholes' device had a piece of printer's type mounted on a little rod, positioned to strike upward to a flat plate (a design used in printing for centuries), which held a piece of carbon paper on some stationary. The strike on the rod, of course, produced an impression on the paper.

Unbelievably, nobody before this time had thought of using the type to strike paper and make an impression on it. Sholes went on to construct a machine to do the whole alphabet. After a long period of trial and error, the prototype was sent to Washington, D.C., for a patent. (At that time the U.S. Patent Office required a working patent model. In fact, the original is locked inside a vault in the Smithsonian today. Today, a working model is not required from the Patent Office.)

Some of the drawbacks to the device were simply challenges to its inventor. One was that the original "up-strike" design prevented the user from seeing what he or she was typing because the impression was made on the paper on the bottom of a cylinder. However, after the design was changed (the new version looked very similar like a modern typewriter) and the user could easily see what was being typed, another problem surfaced. The "type-bar system" and the "universal" keyboard (where all the letters are placed in the same position) was problematic because keys that were next to each other jammed when used together.

To solve the problem, James Densmore, a business associate, suggested splitting up keys for letters commonly used together to slow down typing. This became today's standard "QWERTY" keyboard. Even with all the problems solved, Sholes still faced another obstacle to make his machine a success: He lacked the patience required to market the new product, so he sold the rights to Densmore. Densmore, in turn, convinced Philo Remington (the gunsmith) to market the device. The first Sholes and Glidden Type Writer was offered for sale in 1874, but, as mentioned earlier, was not an instant success.

However, it evolved. Although the original Sholes and Glidden used the QUERTY keyboard, it typed in capitals only and was sluggish and inefficient. But Remington was not the kind to let sleeping typewriters lie, as it were, so he and his engineers worked hard to improve it.

In 1878, the second machine was introduced. It typed both upper- and lowercase letters and used a "shift key" to go between the two. Besides being much quieter, it also featured an open frame (more like modern typewriters) that allowed easier access to the keys and to the inside of the machine for repair. The new model had great sales and was even said to be used and scrutinized by writer Mark Twain. Although he was the first writer to deliver his script to his publisher in typewritten form, he was said to have had it typed from his handwritten copy. Perhaps, even Twain himself didn't trust the newfangled gadget until later in his career. Either way, the age of the typewriter was born and writing and publishing would never be the same.

Diesel Engine 74

Rudolf Diesel, whose name is attached to the engine he invented, was propelled by sociological rather than pecuniary impulses. At the time, the Industrial Revolution was in full swing, and Diesel's dream was to invent an engine that would free people from many of the laborious processes related to other machinery, including gas engines. He wanted people to determine their own destiny, rather than have the machinery they created determine it for them.

Though German, Diesel was born in Paris as the son of an immigrant leather worker. Then, in 1870 the Franco-Prussian War erupted and Diesel and the rest of his family were deported to London as undesirable aliens.

But this did not prevent him from pursuing a technical education. He went to Munich, Germany, where he specialized in thermal engineering and designing machinery. Following his education, Diesel returned to Paris.

Diesel's mechanical goal evolved into a quest to create an engine that was more powerful and functional than the gas engines currently in use. In the gas engine, the spark that ignited the gas, which powered the cylinders, was an external spray, such as a heated filament or electric spark. One of the improvements in his engine was to ignite the fuel internally by having the engine compress the air-fuel mixture in the driving cylinder. As the compression increased, the mixture's temperature also increased to the point that eventually it would combust itself. No firing mechanism would have to ignite it.

Diesel eventually succeeded with this, although his first experimental test engine almost resulted in his demise. The engine exploded and injured him.

Another improvement in the engine was that it ran on low-cost fuel. Originally, Diesel conceived it being powered by coal dust or even animal fat, but he ultimately settled on low-cost crude oil (also called "diesel fuel"). This also resulted in men having to be less involved with the engines of the time.

Diesel's first successful engine ran for about a minute. As he made improvements on it, the demand for it by industry increased. For one thing, it could be built very large, and as such was competitive with the steam engine, which powered many large machines of the time. The diesel engine was cheaper to operate than most other engines because it used less expensive fuel and it cost less to repair. Also, unlike other engines, the diesel engine did not need a long warming-up period to work or a large water supply, like the steam engine.

Diesel patented his engine in 1892—though it could hardly be said at that

Patent drawing, 1898, by Rudolf Diesel. *U.S. Patent Office*

time to be ready for widespread use—but in a relatively short amount of time it was used in a variety of ways in industry where a heavy-duty source of power was required. Later, the engine was also used in tractors, trucks, buses, ships, as well as locomotives and submarines.

There are actually two types or classes of diesel engines. One is the two-stroke, or two-cycle type, where there is a complete cycle of operation in every two strokes of a piston. It needs compressed air for starting as well as operating. In the other type, the four-stroke, or four-cycle engine, the first downstroke of the engine draws in air, and on the upstroke air is compressed to about 500 pounds per square inch. At the top of the stroke, a jet of fuel is sprayed in through an injector, the fuel is ignited, and the rapid expansion of the gas created by the ignited fuel forces the piston down in the firing or working stroke. The next upstroke drives the waste gases out through the exhaust valve and completes the cycle. The amount of fuel injected controls the speed and power of the diesel engine and is not related to the amount of air admitted, as it is in the gasoline engine.

The diesel engine has been around for well over a hundred years, and it is still more economical to use in many instances than the gasoline-powered engine. It is not without flaws, however. It is noisy, and the gases it emits are serious pollutants.

Diesel retained his idealistic attitude his entire life, always trying with all his being to find a place for the individual in an industrial society. His lack of capitalistic savvy was indicated by the fact that though he became a millionaire through his inventions, he did not oversee his investments, so he found himself in constant fiscal difficulty.

Diesel's life ended sadly. His fiscal situation gradually evolved into very bad times and he carried the added burden of worrying about the movement toward war in Europe. Finally, it became too much to bear. In 1913, he vanished without a trace while crossing the English Channel on a night ferry to attend an engineering congress meeting. There is little doubt that he killed himself.

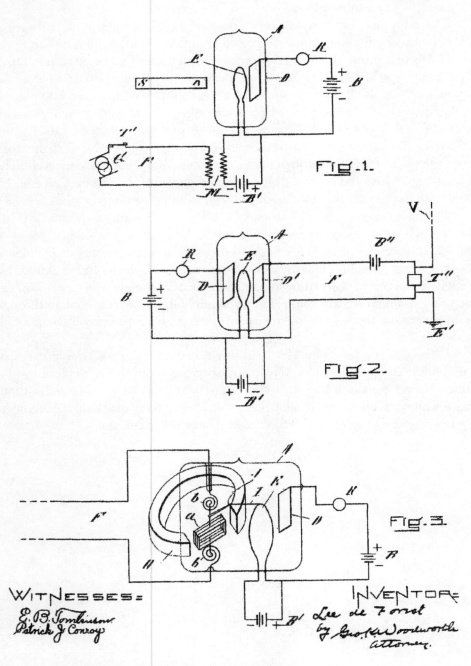

Patent drawing, 1907, by Lee De Forest. *U.S. Patent Office*

Triode Vacuum Tube   75

One day in 1880 when he was in his Menlo Park, New Jersey, laboratory, Thomas Alva Edison attempted to clean carbon off the inside of a light bulb by sticking a piece of wire inside it and was surprised to see current flow from the hot filament in the bulb to the wire. The current had passed through a vacuum. Edison patented the phenomenon, and it—the passing of a current through a vacuum—became known as the "Edison effect."

This was but one of the phenomena that American inventor Lee De Forest was aware of when he succeeded in inventing the triode vacuum tube, one of the most important inventions of the twentieth century because it was able to boost or amplify the signals passing through the transcontinental telephone, radio, and television so that they reached their destinations. For many years, it was to be a vital component of long-distance communications.

De Forest was born in Council Bluffs, Iowa, on August 26, 1873. His father, who was a minister, moved the family to Talladega, Alabama. He importuned his son to study the classics.

Fortunately for the world, De Forest was able to show his father that he had more of an aptitude for science than literature, so his father agreed that he should go to the Sheffield Scientific School at Yale University (his father's alma mater), following a year of prep school in Massachusetts. De Forest took courses in physics and electricity and graduated in 1898. Following graduation, he worked for Western Electric Company and also edited the publication *Western Electrician.*

He was very familiar with how telegraph signals worked. Indeed, in his spare time from Western Electric, he invented a device called a "sponder" for detecting wireless signals. He and E. W. Smyth patented and marketed the device. Ultimately, the device was not successful, but De Forest became involved in a promotional scheme to use his device at the 1903 International Yacht Race in New York, which gave him a lot of publicity.

In 1902, De Forest got involved with Abraham Whiter, a promotor and stock speculator. Together, they opened up the American De Forest Wireless Telegraph Company. The company prospered for a while, but then De Forest discovered that his partners were involved in financial chicanery, so he divested from the company.

But all was not negative. While at the company, he had been working on his "audion," and when he left he was allowed to keep the rights to it.

The groundwork for De Forest's invention began in 1897 when British physicist Sir Joseph J. Thomson not only discovered electrons, but also found that they carry a negative charge. In 1900, Owen W. Richardson, another Englishman, discovered that when metals are heated they emit electrons. Then in 1904, Sir John A. Fleming invented a device for rectifying high-frequency oscillations. His "rectifier" converted alternating current to direct current.

De Forest was, of course, aware of the Edison effect and that Fleming had used Edison's light bulb with the extra electrode and had turned it into a rectifier that allowed current to flow in only one direction. De Forest took the Fleming device and inserted a zigzag piece of platinum wire, which he called a "grid," between the filament and the metal plate. When he did this, he learned that the Fleming rectifier not only kept its rectifying properties, but that it also became extremely good at amplifying signals. His invention came to be known as a "triode." DeForest received a patent for the triode in 1907 and created the De Forest Radio Telephone Company.

As is the case with many inventions, however, the triode was not accepted right away. However, after a variety of public tests, including experimentation with the device by the U.S. Navy, the navy ordered that many of its ships be equipped with the De Forest radiophone equipment.

De Forest, though honest himself, was bedeviled with larcenous partners, and in 1912 he and a couple of them were tried for fraud. De Forest was acquitted, but his legal troubles brought the end of his involvement with the company. He took a job with the Federal Telegraph Company in San Francisco and while there he kept tinkering with his triode, trying to increase its amplification abilities.

De Forest invented many other things (he received more than three hundred patents), but none was as important as the triode vacuum tube. Used in television, radio, and the transcontinental telephone, it opened the door to the field of electronics, only to fade with the invention of the transistor in 1947.

Ultimately, De Forest moved to Hollywood, California, where he died on June 30, 1961.

AC Induction Motor 76

Though he's hardly as well known as Thomas Alva Edison, Polish-born Nikola Tesla is more important in terms of the way electricity functions in the world. To sum up his accomplishment, it was Tesla who championed alternating current (AC). And it was he who invented the AC induction motor, a kind of motor that is used to power a wide variety of electrical devices.

Tesla was born on July 9, 1856, in Smiljan, Croatia, which later became part of Yugoslavia. He had always wanted to be an engineer of some sort, while his father, a member of the Greek Orthodox clergy, wanted Tesla to follow in his footsteps. In those days and in that part of the world, parents were all-powerful and Tesla might well have acquiesced if it hadn't been for his getting sick. Tesla contracted cholera when he was eighteen years old, a disease that kept him in bed for 9 months, an ordeal for anyone but particularly for a young man.

Concerned about Nikola's health and wanting to raise his spirits, he relented to Nikola's desire to be an engineer. Tesla enrolled in the Polytechnic Institute in Grataz, Austria, to study engineering, then went on to complete his studies at the University of Prague in 1880.

It was during his schooldays that he discovered the AC principle and the rotating magnetic field. At the time, the direct current (DC) motor was already in existence, and pioneering work had been done by a variety of inventors on what would be an AC motor. While a student, Tesla became intrigued with the idea of an AC induction motor, and after graduation he worked on developing one. He completed his first model in 1883. Unlike DC, which needed a direct connection to the armature of the motor, AC did not. It created a rotating magnetic field that could drive the motor. AC involved higher voltages than DC, and was regarded by many people as unsafe, but this certainly was not Tesla's belief.

In Europe, Tesla met Charles Bachelor, a close friend of Edison's, who at the time was operating his laboratory in Menlo Park, New Jersey. Bachelor was knowledgeable about matters electrical and was an executive at the Continental Edison Company. Tesla wanted to go to America, so Bachelor wrote him a letter of introduction to Edison.

When Tesla arrived in America in 1884, he was dead broke—indeed, the story goes that he had just 4 cents in his pocket—because he had somehow lost

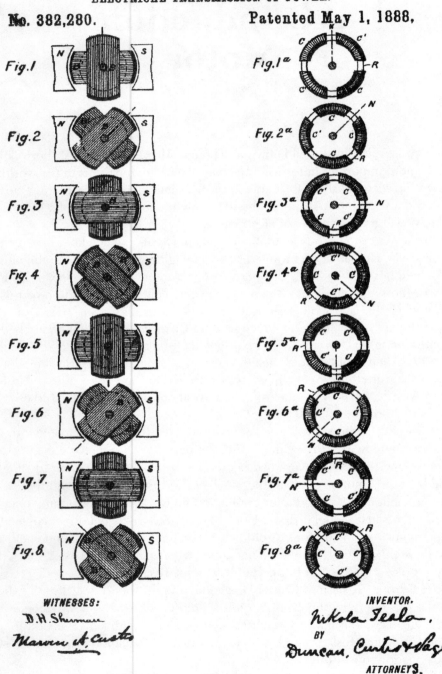

Patent drawing, 1888, by Nikola Tesla. *U.S. Patent Office*

his wallet on the ship during the voyage. But a happy coincidence occurred. As he strolled along Broadway, he came across a group of workmen trying to fix an electric motor. He stopped, examined the motor, and fixed it, for which the group gave him $20, a munificent sum in the 1880s (where a bottle of milk cost only a few cents).

Tesla presented the letter to Edison, who promptly hired him. But they did not get along that well because of their high-blown egos and because they had a fundamental difference in terms of electricity: Tesla believed in AC, while Edison championed DC. Ultimately, their relationship melted down when Tesla quit over, he said, Edison not paying him the $50,000 he was promised for having improved the Edison generator.

In 1887, he created the Tesla Electric Company with the financial backing of others and started to manufacture his AC induction motor. On October 12, 1887, he filed a patent application that covered polyphase and single motors as the distribution system and transformers. In the end, his application was broken down into seven separate inventions, for which he was given patents.

Tesla's achievements did not go unnoticed, as George Westinghouse, a fellow inventor and businessman, bought his patents for $1 million—a king's ransom at the time. Tesla worked for Westinghouse in Pittsburgh, Pennsylvania. In early 1889, Westinghouse's company offered consumers a fan powered by a $\frac{1}{6}$-horsepower (hp) AC motor.

The battle over which was better—DC or AC—continued until Westinghouse won a decisive victory over Edison when his company was given a contract to build a large generating plant at Niagara Falls using, of course, AC. This victory was won primarily because AC had proved its worth at the World Columbian Exhibition in Chicago where AC generators provided power for the lights. The plant was erected in 1896 and featured three Tesla AC dynamos, each generating 5000 hp.

Eventually, the eccentric Tesla separated from Westinghouse. Though he continued to be regarded as the transcendent electrical genius that he was, many of his later experiments failed and of the few successful inventions he made, he neglected to obtain patents. He died on January 7, 1943, while subsisting on a small pension from the Polish government.

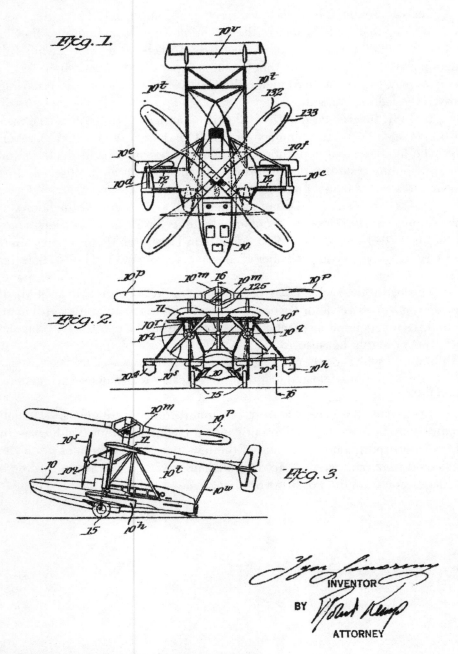

Patent drawing, 1929, by Igor Sikorsky. *U.S. Patent Office*

Helicopter 77

One of the ideas emerging from the fecund mind of Leonardo da Vinci and found among his drawings was a helicopterlike machine, so it is commonly assumed that Leonardo was the first person to conceive of such a contraption. In fact, there is evidence that Chinese and Renaissance Europeans had the design in mind, because among the artifacts found from these civilizations are toys that look like helicopters.

Various inventors had tried to make a working helicopter, but the problem was usually the same: finding an engine that could make a "blade" whirl with enough power to create the "lift," or vertical thrust, to get the craft off the ground.

In 1907, a helicopter, designed by Paul Cornu, was able to get up off the ground. In 1923, a Spaniard named Juan de la Cierva successfully flew an "autogiro," but it wasn't until the 1930s and the work Igor Sikorsky, whose name has become synonymous with helicopters, that a practical craft was developed.

Sikorsky was born in Kiev, Russia, on May 25, 1889, the youngest of five children. His mother and father were prominent personages in the tsarist regime, his father a professor of psychology at St. Vladimir University in Kiev, his mother a graduate of a medical school. They were closely allied with the tsar and lived a life befitting royalty.

When a young boy, Sikorsky became interested in all of Leonardo's aeronautical drawings, particularly the helicopter, and he started on an education that was focused on an aeronautical career. While a teenager, Sikorsky studied in Germany, and then traveled to Paris, the mecca of aeronautical studies, to study aeronautical design concepts.

It was at this time that Sikorsky also imagined building a practical helicopter, and while in Paris he bought a 25 horsepower (hp) engine to power a single-blade design he had created. But the invention was bedeviled by the same problem that other designers had run into: The engine wasn't powerful enough to provide the vertical thrust to get his craft off the ground.

Sikorsky dropped his experiments for a while, designing various fixed-wing aircraft, including a number of award-winning military craft, such as bombers, for the Tsarist Imperial Army. As such, Sikorsky was well identified with the tsar. When the communists came to power after the revolution, he was one of the people marked by them for imprisonment or worse, so he fled the country, giving up his aeronautical career and land holdings in Russia.

He ended up in France, where he was commissioned to build a bomber for the Allies who were still fighting World War I, but he never got to complete it. It was still on the drawing board when the armistice was signed in 1918 and the French canceled the contract. In 1919, he left France and when he arrived in New York City he was virtually penniless and had to live an impoverished lifestyle.

For the next ten years, starting in the early 1920s when he obtained financial backing to open his own company, Sikorsky Aero Engineering Corporation, on a farm near Roosevelt Field, Long Island, Sikorsky developed fixed-wing airplanes. It was only in the 1930s that he eventually came back to his original dream of designing a flying helicopter.

To develop his helicopter, he needed money, so he applied to United Aircraft, which invested around $300,000. On September 14, 1939, Sikorsky climbed into what was truly the first single-rotor helicopter, a collection of pipes welded together, an open-air cockpit, and a three-bladed rotor powered by a 75 hp engine that turned an automobile fan belt that turned the blades.

The assemblage flew, and Sikorsky was thrilled. Later on, he said, "It is a dream to feel the machine lift you gently up in the air, float smoothly over one spot for indefinite periods, move up or down under good control, as well as not only forward or backward but up in any direction."

Sikorsky dubbed this first helicopter, the "VS-300," and it came on the scene as World War II broke out. The U.S. Army ordered a variation of the VS-300 called the "R-4."

The helicopter was not greatly used in World War II, but it came into its own during the Korean War, which started in 1950. Because it could land where no other aircraft could, it became an essential, used by the U.S. Army in a variety of ways, including observation, transporting wounded, and hauling important cargo.

As the years went by, the Sikorsky helicopter became more complex and useful, and served in a variety of roles, such as a "troop carrier" and "gunship." Other significant developments included a "skycrane" helicopter, which could haul up to 20,000 pounds hanging on a cable, and an amphibious helicopter.

Sikorsky was particularly pleased with the helicopter's ability to save lives rather than destroy. He said once, "It was a source of great satisfaction to all the personnel of our organization, including myself, that the helicopter [had a] practical career by saving a number of lives and by helping man in need rather than by spreading death and destruction."

Calculator   78

If time is of value, then the calculator deserves to be on any list of the one hundred greatest inventions of all time. It has saved enormous amounts of time for accountants, storekeepers, and a wide variety of other people who have to make mathematical calculations in their work.

The calculator was a long time in coming. At first, people counted—literally—with their fingers to make calculations. But this ultimately changed into the abacus in China and the soroban in Japan. It wasn't until hundreds of years later, in 1614, that John Napier, a Scottish mathematician used marked strips of bone that were rearranged into fixed positions to make various calculations. "Napier's bones," as his method was called, evolved into the "slide rule." French mathematician Blaise Pascal made the first machine that resembled the modern-day calculator, embodying as it did, wheels.

A variety of other inventors helped to develop the calculator. For example, Thomas de Colmar, who lived in Alsace Lorraine, made a commercial machine in 1820 called the "arithmometer." And Charles Babbage, an Englishman, was working on an automatic calculating machine in 1871 when he died.

The first calculating machine patented in America was by O. L. Castle of Alton, Illinois, in 1850. It had ten keys and only added up numbers in one column. It also did not print.

Another patent for a calculating machine was granted in 1875 to Frank Baldwin, and though the machine did not work that well, Baldwin nevertheless received at least one prestigious award for the device: the John Scott Medal from the Franklin Institute.

Like so many other inventions, such as the light bulb and the steam engine, it took the work of one man, William Seward Burroughs, to perfect the calculator to the point where it would receive wide acceptance.

Burroughs was born in Auburn, New York, on January 28, 1855. His father, Edmund, was a builder of models for new inventions, but at first Burroughs did not get involved in his work. Instead, starting at the age of fifteen, he obtained a job in a bank as a bookkeeper.

Burroughs found much of the work tedious and repetitive, with 90 percent of the calculations having to be done by hand. He wondered if he could invent a device to make all this work simpler and to reduce the number of hours on the

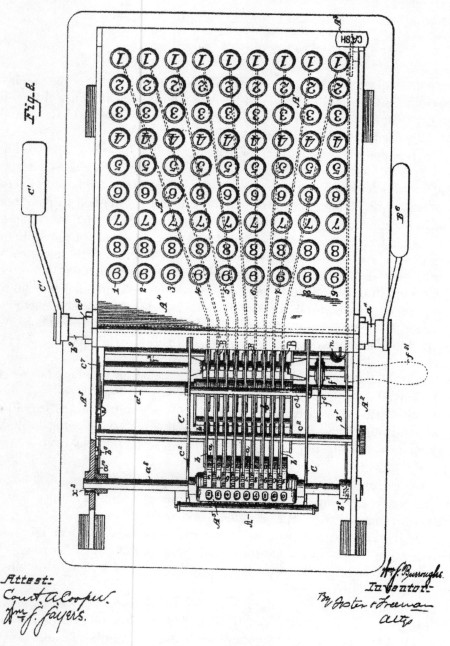

Patent drawing, 1888, by William Seward Burroughs. *U.S. Patent Office*

job, which was affecting his health. But the work load finally got to him, and he was forced to resign.

He and his family moved to St. Louis, and for a while Burroughs worked in his father's model making shop, but he also continued to work on a calculator. Then one day he showed what he was working on to the financier Thomas B. Metcalf, who exhorted him to continue work on the machine. This he started to do in earnest at a machine shop in St. Louis.

By 1885—Burroughs was by then only thirty years old—he had completed a machine that could calculate, record, and print data, and Metcalf and two St. Louis merchants decided to back it, forming the company American Arithometer Company to produce it. Produce it they did, and it was accepted by the public to some degree.

But there was a problem that neither the inventor nor his backers saw, and it was related to the public's use of the machine. To operate the machine, the user had to pull a handle all the way down and then release it to record a transaction. The problem was that there was no way to control the speed at which the handle was pulled down, and if someone did it too fast it would affect the results in a negative way. Gradually, the public, which seemed incapable of learning the right way to pull the handle, stopped buying the machine, and the company started to inch its way toward bankruptcy.

But Burroughs saved the day. In 1890, he invented a handle that did not allow itself to be used incorrectly. It contained a small cylinder, partly filled with oil, and a plunger or rod. When the handle was pulled down, the cylinder only allowed it to go at a certain speed, no matter how quickly the user pulled it. In other words, it acted as a shock absorber. A few other refinements were added—no pun intended—to the machine and then it was offered for sale.

Sales were on the slow side in 1894 when only 284 machines were sold, but this gradually increased and by 1904 1,000 machines were sold. By 1913, the company was doing annual sales of $8 million, a large sum at that time.

Unfortunately, Burroughs was not around to witness the complete fruit of his handiwork. After a lifetime of poor health, he was eventually diagnosed with tuberculosis. He died on September 14, 1898. But the company that he helped create is still in existence.

79

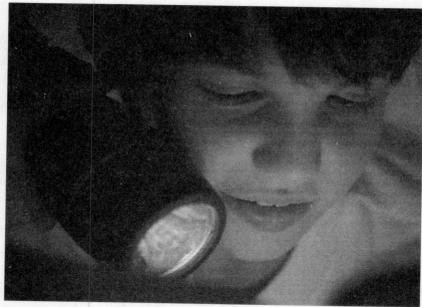

The flashlight is another underappreciated invention—until you need it. *Duracell*

Flashlight

If you needed to find something in a dark area of the house before 1896, it could be hazardous. You had to use a candle or a lamp filled with whale or motor oil, and fires were not uncommon.

The first trustworthy, portable safe lighting device was the flashlight, invented in 1896. The electric lights were called "flashlights" since they did not give off a long, steady stream of light. Instead, crude switches were used to turn them on for very brief periods in flashes of light. This was because the bulbs at the time were inefficient and the batteries were weak.

The evolution of the flashlight is closely associated with the development of batteries and bulbs. Georges Leclanché, a French inventor, invented the battery in 1866. He called it a single-fluid electric battery. Known as a "wet cell," it was extremely impractical, consisting, as it did, of a glass jar filled with ammonium chloride, manganese dioxide, and zinc that was prone to tipping over. A carbon bar was inserted in one end that acted like the positive end of the cell.

Improvement of the battery came in 1888 when Carl Gassner, a German scientist, encased the wet-cell chemicals in a sealed zinc container. This made the battery a "dry cell" because the contents were encased and the outside of the battery remained dry. To this day, batteries are made this way.

Bulbs were another story. Thomas Alva Edison invented the light bulb. At the time, the carbon filament in the bulb was not very efficient and the glow of a bulb was sometimes sporadic, hence "flashlight."

The first tubular flashlight was invented by David Misell, who also invented the early bicycle light. In 1895, a number 6 battery was needed to produce enough light, but it was 6 inches high and weighed over 3 pounds. A year later, the "D-cell" battery was invented and several of them together could produce the power of the number 6, thus making, for the first time, a practical, hand-held, portable light.

Patented on November 15, 1898, the O. T. Bugg Friendly Beacon Electric Candle was sold by the U.S. Battery Company. It was 8 inches long and had two D-cell batteries encased in an upright tube, with the bulb protruding from the middle of the cylinder.

An invention in 1906 made the flashlight light even brighter. The tungsten wire filament replaced the carbon filament that Edison had used to make the light bulb a reality. About this same time, improved switches began to appear on flashlights and the length of time they could work was increased. Another improvement in 1911 replaced the "push-button" switch with a simple "slide" switch that was easier to operate, particularly with one hand.

So, Misell received word back from the field that the flashlights worked. By about 1897, he had patented several flashlights. When his next patent was granted on April 26, 1898, it was assigned to Conrad Hubert's company, a friend and coworker of Misell's. Hubert's company, the American Electrical Novelty and Manufacturing Company, later became Eveready.

Growth continued and by 1899 a catalog for the company featured twenty-five different bulbs and batteries. By 1902, the name Eveready was listed in the advertisements. By 1924, Eveready introduced the "safety lock" switch, which was wider and flatter then previous ones and combined the button- and slide-switch styles. This style was used into the 1930s.

Interestingly, the flashlight also took part in the development of the atomic bomb. The first nuclear reaction, as noted elsewhere in this book, was done under the stands at Stagg Field at the University of Chicago. The first atomic reactor was immense: 30 feet wide, 32 feet long, and 21 feet high, weighed 1,400 tons, and contained 52 tons of uranium. But on its maiden effort, it only produced enough energy to make a small flashlight work.

Today, flashlights are an integral part of our daily life, necessary to have nearby in case the lights go out, in the trunk of our cars in case of mechanical problems, and on camping trips. In some professions, the larger flashlight is considered a required piece of equipment. Police, for example, are trained in its uses, not only when investigating something at night, but also as a defensive or offensive weapon.

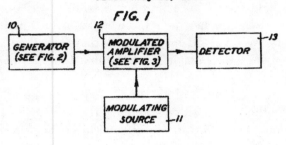
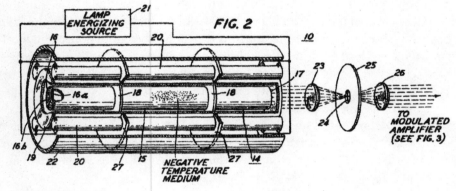
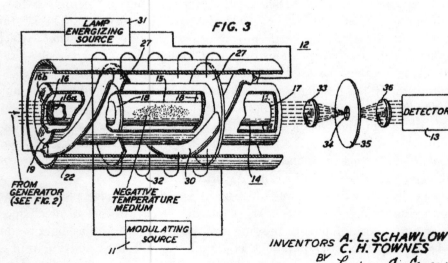

Patent drawing, 1958, by Arthur L. Schawlow and Charles H. Townes. *U.S. Patent Office*

Laser   80

The laser's invention dates to 1958 with publication in *Physical Review* of "Infrared and Optical Masers" by Arthur L. Schawlow and Charles H. Townes, both physicists employed by Bell Labs. A new field, leading to a multibillion-dollar industry, had gotten its start.

Schawlow and Townes started out in the 1940s and 1950s interested in the field of microwave spectroscopy and in exploring different features of molecules. They weren't planning to invent anything to revolutionize communications or medicine; they just tried to develop something for studying molecular structures.

Townes, with a doctorate in physics from the California Institute of Technology, joined Bell Labs in 1939, where he spent his time with vacuum tubes, microwave generation and magnetics, solid-state physics, and electron emissions from surfaces. Soon after he arrived, his lab group was told to start developing a radar navigation bombing system. World War II would be fought in science's halls as well as on European and Asian battlefields.

Though interested in radio astronomy, Townes worked on radar, which kept his focus on microwave spectroscopy. (Radar broadcasts radio signals at specific wavelengths that strike solid objects, like battleships or airplanes, and reflect back to the radar system where they can be correlated to identify the object and its position.)

Townes' radar navigation bombing system used wavelengths of 10 centimeters (cm) and 3 cm. The military wanted 1¼ cm so its planes could use smaller antennae. Townes was uncertain about this: He knew that gas molecules absorb wave forms at certain wavelengths, which might mean that atmospheric water vapor—such as fog, rain, and clouds—would absorb smaller radar signals. His concern proved well founded: water vapor did interfere with the signals.

In 1948, after the war, Townes left Bell Labs for Columbia University. Columbia was more interested in the principles of physics that intrigued him. And he liked university life. Schawlow, with a doctorate in physics from the University of Toronto, went to Columbia on a fellowship. He began working with Townes in 1949. Townes continued mulling over the possibility of using stimulated emissions to probe different gasses for molecular spectroscopy. This challenge led to the invention of the "maser" (microwave amplification by stimulated emission of radiation), then the "laser" (light amplification by stimulated emission of radiation).

Townes understood that as the wavelength of microwave radiation grew shorter, its interactions with molecules became stronger, making it a more powerful spectroscopic tool. But building a device that could generate the required wavelength was beyond fabrication techniques of the day. Townes eventually developed the idea of using molecules to generate the desired frequencies.

Various technical issues had to be resolved, however, like getting around the second law of thermodynamics. That law basically implies molecules cannot generate more than a certain amount of energy, but how to get around this came to him while attending a conference. During an early morning walk in Franklin Park (in Washington, D.C.), Townes mulled over his problem. "But I thought, now wait a minute! The Second Law of Thermodynamics assumes thermal equilibrium. We don't have to have that!" He pulled an envelope out of his jacket and started to jot down a calculation on how many molecules he needed in a "resonator" to get the power output he wanted. Back at his hotel he told Schawlow, who was also in Washington for a conference, about his idea. "I talked to him about it and Art immediately saw the point and said, 'That's interesting.'"

When Townes returned to Columbia, he asked James P. Gordon, a graduate student, if he wanted to work on the project, "I think it is going to work, but I can't be sure." He later hired H. L. Zeiger to help. Schawlow didn't work on the maser, but said, "I had witnessed the disclosure in his notebook."

Later that year, Schawlow left Columbia, for a research job at Bell Labs. He worked on superconductivity, so he was not part of maser development during the ensuing years.

Townes decided to use ammonia, a strong absorber that interacts with wavelengths. It was an "old favorite" and he knew a lot about it; he had cavities at the necessary wavelength (1¼ cm), the techniques, and the wave guides. Not too many others seemed interested in his maser idea, so he bumped along "in a graduate student kind of style." Within three years, he had made it work.

In 1953, he, Gordon, and Zeiger demonstrated a working device that he called the "maser" (microwave amplification by stimulated emission of radiation). It was patented through Columbia. Townes decided that wavelengths shorter than microwaves (infrared and optical light) might be better for spectroscopy than the wavelengths produced by the maser.

In 1956, Bell Labs offered Townes, who was still at Columbia, a consulting job that would let him visit the labs, talk to people, look at projects, and exchange ideas. "Well, that was a nice kind of consulting job, so I accepted it."

Townes was still thinking about stimulating the emission of light. At Bell Labs, he stopped in to see Schawlow (now his brother-in-law), who had been there around five years. They had collaborated before by coauthoring the 1955 book *Microwave Spectroscopy*. Schawlow said of their work later, "I was beginning to think seriously about the possibility of extending the maser principle from the microwave region to shorter wavelengths, such as the infrared region of the spectrum. It turned out that he was also thinking about this problem, so we decided to look at the problem together."

Schawlow wanted to arrange a set of mirrors, one at each end of the cavity, to bounce the light back and forth to eliminate amplifying beams bouncing in other directions. Schawlow and Townes talked this over and started working on the principles of a device capable of providing these shorter wavelengths. Schawlow thought that the mirrors' dimensions could be adjusted so the laser would have only one frequency. A particular frequency could be chosen within a line width, and mirror size could be adjusted so that any off-angle motion could be dampened. The net result was elimination of most of the cavity, while keeping only the two ends.

They worked on it intermittently over the next few months. Schawlow worked on the device, Townes on the theory. Schawlow thought of using some solid-state materials for solid-state lasers. They had not yet made an actual laser but collaborated in 1958, as already mentioned, on a paper that extended the principles of the maser to the optical regions of the spectrum. They applied for a patent through Bell Labs, receiving it in 1960, the same year a working laser was built by Theodore Maiman at Hughes Aircraft Company.

In 1961, Schawlow left Bell Labs to take a teaching and research job at Stanford University (they "made an offer I couldn't refuse," he said), where he developed the laser's use in spectroscopy.

In 1964, Townes shared the Nobel Prize in physics with Aleksandr Prokhorov and Nicolay Basov of the Lebedev Institute in Moscow for "fundamental work in the field of quantum electronics which has led to the construction of oscillators and amplifiers based on the maser-laser principle." In 1981, Schawlow also shared the Nobel Prize in physics for his "contribution to the development of laser spectroscopy." "It's long overdue," Townes said of his colleague's honor at the time.

Schawlow recalled the years spent on invention, "We thought it might have some communications and scientific uses, but we had no application in mind. If we had, it might have hampered us and not worked out as well."

81

Robert Fulton designed the first efficient steamboat.
New York Public Library Picture Collection

Steamboat

Many people think that Robert Fulton invented the steamboat. In fact, variations of a steam-powered boat were on the scene (and in the water) before Fulton. His achievement was to show that a properly made steamboat was technically achievable and could be a viable mode of transportation. In a very real sense, then, he did invent it, or is considered by many as the inventor.

Fulton was the son of Robert Fulton, a leading citizen of Lancaster County, Pennsylvania. Early on, Fulton showed a remarkable ability to draw and when just a teenager he was employed by local gunsmiths to draw designs for their guns.

His abilities were so good that when he was seventeen he left Lancaster County for Philadelphia to establish himself as a portrait painter and miniaturist (an artist who painted miniature pictures on cameos and the like). Four years later, he decided to advance his painting education, so he left America for England intending to study under the painter Benjamin West.

But he was not prepared for what greeted him in England. The Industrial Revolution was in full swing, with canals, factories, mines, bridges, and all kinds of new equipment coming into play. Fulton was enthralled with it all, so much so that he switched careers, putting aside painting for engineering. It was a wise and happy choice for humanity.

When he was just fourteen years old, Fulton had designed a steamboat powered by a paddle wheel. Now he wanted to put his design into action. He appealed to the British government to let him buy a steam engine and ship it to America. The government, having put a ban on this practice, refused his request. Why it refused is not known, but one historian speculates that it was simply because Fulton was an American, an Irish American at that.

Fulton kept trying for three years, during which he was by no means idle. He designed and patented a variety of devices, including one for hauling canal boats over difficult terrain, dredges for canal work, and a machine that could twist hemp into rope. But the British government was adamant about not shipping the engine, so after a while Fulton gave up and went to France, where he tried unsuccessfully to interest people in his inventions. He also conducted tests on a submarine, called the *Nautilus*, that he had invented as well as fitting steam engines on ships.

Then an event occurred that eased the way. He met Robert Livingston, the American minister to the French government and who had been a partner in another steamboat invention. In 1803, the two men importuned the British government again, and this time they were able to buy an engine from the firm of Boulton and Watt. Still, they had to wait another three years before they could get authorization to ship it to the United States. Once in New York City, Fulton and Livingston set about the task of making the idea work. Livingston favored a rear paddle, while Fulton favored a paddle wheel on each side, the design they eventually chose

They installed a 24-horsepower steam engine into the 100-foot-long *Clermont*, and on August 17, 1807, it made its first voyage up the Hudson River, traveling about 5 miles per hour. The trip was successful, and a few weeks later the boat made its first commercial run. Only a few hardy souls boarded the ship but as the *Clermont* kept to its regular schedule, more and more passengers used it. By the time the operation was closed for the winter, a small profit was being turned.

Because it could give such a smooth ride (but only if the water was smooth, as on the Hudson or Mississippi Rivers), Fulton was able to install furniture, and thus he introduced Americans to luxury steamboating. He built twenty more boats, each fancier than its predecessor, and steamboats spread across the country.

When the War of 1812 erupted, Fulton focused on designing submarines and warships. He was in the midst of building a huge steam-powered ship with great destructive capacity when he died from a respiratory ailment on February 24, 1815, shortly after news of the war's end reached America from Belgium, where the Treaty of Ghent was signed.

Steam-powered boats prospered for a long time, and were only displaced as other fuels came into place for boats. Regardless, steamboat passengers had to be hardy souls to ride steamboats, because there were quite a few spectacular accidents where engines blew up and resulted in a great loss of life. In general though, for its time, the steamboat performed quite well.

82

Modern fax machine with phone. *Author*

Fax Machine

In recent years, the facsimile (fax) machine has been supplanted, to some degree, by the sending and receiving of messages via computer. But it seems quite certain that it will be around for a long time to come.

The fax machine, which essentially involves sending images electrically, was patented 1843, though it only came into wide use relatively recently. The heart of the idea is related to the discovery by French physicist Alexandre-Edmond Becquerel, that when two pieces of metal are immersed in an electrolyte and one of them is illuminated, an electric charge develops. In essence, he had discovered what the electromechanical effects of light were, but Becquerel had no idea how they might be put into practical use.

Alexander Bain's idea was to electrify raised metal letters. Then, using a stylus that was secured to a pendulum, he outlined or followed the shape of the letters as the pendulum moved slightly on each passage. The generated currents would then pass across a telegraph line, and when the currents passed to a synchronized pendulum that was in contact with paper that had been soaked in potassium iodine, the letters appeared in light brown configurations.

The idea was quite clever, but as depicted in the patent application there had to be exact synchronization at both ends for the device to work. To do this, the patent shows two identical telegraph devices, one for sending and the other for receiving, that have built-in magnets at the top and coils of insulated wire between them.

Further progress was made on the fax when Frederick Blakewell, a physicist in Middlesex, England, used tinfoil wrapped around synchronized rotating cylinders instead of metal, which allowed drawings to be sent. A model of this machine was first demonstrated at the Great Exhibition of 1851.

The next step on the fax parade was the work of Giovanni Caselli. Born in Siena, Italy, Caselli was a priest before he joined the revolutionary activities in Italy, which forced him to flee to Florence. There, he taught physics and worked on developing a device he called a "pantelegraph," which used the ideas from the faxes invented by Blakewell and Bain. To use the pantelegraph, the user wrote a message in nonconducting ink on a sheet of tin. The tin was then secured to a curved metal plate and scanned by a needle, three lines for every millimeter. The signals to the receiving machine were communicated by telegraph, and the message was written in Prussian blue ink because the receiver paper had been bathed in potassium ferrocyanide (ferrocyanide is used in making blue pigments). Absolutely accurate synchronization of the needles at both the sending and receiving ends was crucial. To ensure that this device was properly equipped, he installed accurate clocks that worked in tandem and that would activate a pendulum that was linked to a series of pulleys and gears to which the needles were attached.

The French government became very interested in Caselli's fax and had him conduct a series of tests to explore further interest. The machine passed; a working fax was installed between Paris and Lyons in 1865, and later was extended to Marseilles to transport commercial information, such as stock prices, but drawings were also sent.

The machine might have been in much wider use much earlier but its development was interrupted by the Franco-Prussian War in 1870; attention was diverted away from it and the service was never resumed. Then, in 1891 Caselli died in Florence without doing further work on his machine.

Bain, one of the prime movers of the fax, had a sad ending. Originally, he had been awarded 7,000 pounds by the British government for his groundbreaking work in telegraphic machines, but was then threatened with litigation and lost the entire 7,000 pounds. In 1873, a group of inventors implored Prime Minister William Gladstone to award Bain with a yearly stipend of 80 pounds, which he did. Regardless, Bain died in obscurity in a town near Glasgow in 1877.

83

U.S. Army light tanks, circa World War II. *Photofest*

Tank

As a military weapon, the tank forever changed the face of battlefields. It came about as an answer to a practical need. The first modern military tank was developed by the British and French during World War I as a weapon to penetrate barbed wire entanglements and overwhelm enemy machine-gun nests and bunkers.

However, the beginnings of the tank reach as far back as the 1770s. No one individual can be credited with inventing the tank. Instead, a number of gradual technological developments such as the steam engine and the internal combustion engine, brought about the development of the tank as we know it.

The first tanks were really steam-powered tractors that could traverse muddy terrain. During the Crimean War, John Edgework created a better "caterpillar" track that allowed the tank to go places it previously couldn't reach. Still, it wasn't until 1885 that it began to be a viable machine with the invention of the internal combustion engine. The tank, then, did not require huge amounts of water to convert to steam. Just fill up the gas tank and away it went.

In 1899, Frederick Simms designed what he termed a "motor-war car." It had a powerful engine, bullet-proof plating, and two revolving machine guns. He offered it to the British government, but it didn't think the invention was that useful.

But a good idea doesn't usually go away. At one point, the company Killen-Strait developed a tank that featured an improved track, which was made of steel pins and links that meshed together. Later, Hornsby and Sons produced the Killen-Strait Armoured Tractor. The caterpillar track was improved with steel links meshed together with steel pins.

When World War I started, the tractor was again shown to the British government and military officials, and it clearly showed how easily it could plow through a barbed wire fence. Indeed, at the demonstration was a young man named Winston Churchill, who liked it very much and appointed a committee to see how significant it was for warfare. The testing of the tractor was conducted in utmost secrecy, and it was dubbed "tank" because it looked like a water carrier.

The first tank, nicknamed *Little Willie* weighed 14 tons (more than the average elephant), was 12 feet long, and had a crew of 3. It had a snail-like speed of 3 miles per hour on flat ground and dropped to 2 miles per hour on rough terrain. Originally, it could not cross trenches but this was eventually changed.

The first tanks were difficult to navigate. They were hot and cramped inside, and experienced frequent breakdowns during battlefield conditions. Indeed, due to their weight tanks easily became bogged down in mud and had to be towed out by other tanks or be dug out by hand.

However, the tank met and passed its first real battlefield test when the entire British Tank Corps (consisting of 474 tanks) saw action in the Battle of Cambrai on November 20, 1917. British troops gained advantage when the tank corps breached 12 miles of the German front, which resulted in the capture of 10,000 German soldiers, 123 artillary guns, and 281 machine guns. Although Britain's initial success was later cancelled out by Germany's counterattacks, the success restored diminishing faith in the tank and started the Germans thinking about its efficacy.

Increasingly, tanks were used during the Allied advance in the summer of 1918. Tank deployment on a grand scale was reached on August 8, 1918, when 604 Allied tanks assisted in a 20-mile advance on the front.

By the time the war came to a close, the British had produced 2,636 tanks. The French produced 3,870. The Germans, never fully convinced of the tank's benefits, and despite the record of German technological innovation, produced only twenty.

Many of the improvements made on modern tanks include more maneuverability, more comfortable cockpits, less heat and noise, and, of course, an increase in fire power. Today's tanks are computer driven and have all the latest technological devices the military can spare. Some of these include computer-assisted navigational systems.

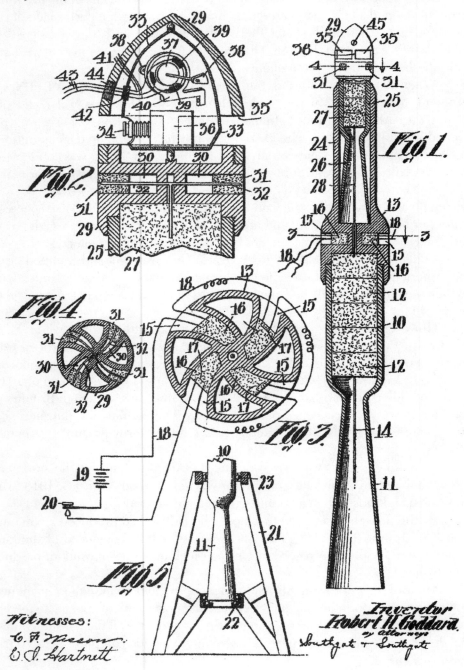

Patent drawing, 1914, by Robert H. Goddard. *U.S. Patent Office*

Rocket 84

Rockets were not a development of the twentieth century. Records dating to ancient China show how potassium nitrate, sulfur, and charcoal—early gunpowder—were stuffed into bamboo tubes and launched into the air. The Chinese went on to develop multistage skyrockets that included color flares, much as the latest creation of the famous fireworks family, the Gruccis. This technology, unfortunately, evolved into weapons of war: the cannon and the rifle. Space flight as well was envisioned as a concurrent dream of practical flight in our own skies. By the late nineteenth century, Jules Verne, in his epic science-fiction tale *From the Earth to the Moon,* used a large cannon to blast his brave voyagers into space in a large artillery shell! Pure fantasy!

Many people, however, took the prospect of space flight seriously and early on came to realize that solid-fuel "gunpowder" missiles would never suffice. Space flight required high-thrust sustainability over long periods of time. Guiding the missile might necessitate turning the motor on or off at selected intervals. And, solid fuels had an inadequate pound-for-pound power-to-weight ratio. This is of critical importance in overcoming Earth's gravitational field to achieve "escape velocity." Even the earliest successful liquid-fueled motors produced over a half-million horsepower of thrust while weighing less than half a ton!

Many people experimented with rockets. Important—and widely published work—was done by the Russian mathematician Konstantin Tsiolkovsky. Born in 1857 and self-educated, he became fascinated with the prospects of interplanetary flight. By the 1890s, he had produced mathematical formulas governing rocket motion. In 1903, he published his work, which included not only his mathematical theories, but also far-reaching proposals for space satellites, space suits, and even showers for the astronauts to be used in the weightless environment of outer space. Tsiolkovsky also made explicit arguments for the use of liquid fuels. He suggested liquid oxygen and hydrogen, which were to be used in rocket experiments years later.

As is often the case, most of Tsiolkovsky's work was originally ignored or ridiculed. Later, he was hailed by the Soviets as the true inventor of the rocket and was given a state funeral when he died in 1935.

"Every liquid-fueled rocket that flies is a Goddard rocket," wrote rocket scientist Jerome Hunsaker. Robert Hutchings Goddard was to take center stage in rocket development in the 1920s, and is today credited by nearly all as the Father of the Space Age.

Born in Worcester, Massachusetts, in 1882, Goddard grew up as a sickly child. Often confined to bed, he read such novels as H. G. Wells's *War of the Worlds* and became enchanted with thoughts of rockets and outer space. While a physics major at Clark University, he began preliminary research into rocket propulsion. At first, his work went along the lines of solid fuel, but he came to the conclusion that only a liquid-fueled rocket would meet the requirements for space flight. His choice of fuel—liquid oxygen and hydrogen—was arrived at independently of Tsiolkovsky.

After receiving his doctorate in 1911, Goddard became a teacher at Clark and interspersed his lectures with advanced proposals on explorations of the Moon and nearby planets with the use of rockets. Moving on to practical work, Goddard obtained a grant from the Smithsonian Institution in 1916 and began to design and build actual rockets. World War I interrupted this pursuit, but Goddard designed a solid-fueled rocket—a predecessor of the bazooka—for the Army Signal Corps.

Continuing with his liquid-fueled experiments after the war, Goddard soon became the object of ridicule. One newspaper dubbed him the "Moon-rocket Man" and the *New York Times* stated that he had not even "the knowledge daily ladled out in high school." The Smithsonian, although agreeing to subsidize him, became restless with his apparent lack of progress.

Undaunted, Goddard kept up his work. In 1929, Charles Lindbergh—then America's greatest hero—took an interest in Goddard's work. Lindbergh convinced financier Daniel Guggenheim to underwrite a $50,000 grant. Goddard was then able to relocate his test to Roswell, New Mexico, at a place called Eden Valley.

At Roswell, Goddard made significant progress. Foremost was a gyroscopic stabilizing device that soon had his rockets getting off the ground and into the air in a predictable straight path. He also devised a parachute recovery system to recover rockets, a method still in use today. By the late 1930s—with his small staff, (which included his wife as the official photographer), mail order, and scrap parts—Goddard was achieving regular, perfect launches of over 7,000 feet.

With war in Europe approaching, Goddard tried to interest U.S. military authorities in his inventions. They, unfortunately, were not very interested. Regardless, Goddard was justifiably concerned about progress in Nazi Germany. There had been rocket societies in America and Britain since the 1920s. They had produced some interesting experimental paraphernalia, though nothing on Goddard's level. Germany, however, was another story.

When the Nazis came to power in 1933, they wasted no time in appreciating the rocket as a potential military weapon. The society was quickly put under the control of army officials. Wernher von Braun, a young student barely out of his teens, had demonstrated a keen ability for design and organization, so he was put in charge of rocket experimentation.

Von Braun and his staff developed the line of "A series" rocket engines, which gradually increased in sophistication and size. Much of his work was close to Goddard's.

By September, 1944, the V2 "terror bomb" was ready for deployment against the British Isles. The Germans had been attacking Britain, beginning in June 1944, with the V1 "Buzz Bomb," a jet-powered drone similar to an airplane. However frightening, the V1 was slow enough to be intercepted by British artillery and fighter planes. Not so the V2. Forty-six-feet in length and weighing 14 tons, the V2 flew at an altitude of 50 miles and traveled 3,000 miles per hour! Fortunately, the German military collapse prevented further use of either the V2 or other "terror weapons."

Once World War II ended, the Cold War soon took its place. Part of that war focused on the "the race for space." Von Braun defected to America. As an advisor and later head of the U.S. space program, Von Braun began by utilizing salvaged V2s in research experiments. Working at the White Sands proving ground in New Mexico, Von Braun developed a two-stage rocket in 1949. Using the V2 as a "booster" to send a small WAC Corporal rocket 250 miles into the atmosphere, Von Braun predicted that a Moon landing was feasible by the 1960s! Von Braun was able to see his prediction come true several times over before he died in 1977.

Goddard worked at Annapolis during World War II and helped design a solid-fueled assist-take-off unit for naval planes. Before he died in 1945, Goddard was able to examine a captured V2, which astounded him for its size and similarity to his own designs. Goddard had over two hundred patents and liquid-fueled rockets. In 1960, the U.S. government made a special $1 million contribution to his estate to further the use of his ideas for rocket research. The best testimony to his contribution was made by Von Braun, who remarked after examining Goddard's patents in the 1950s, "Goddard was ahead of us all."

85

The cotton gin made a huge difference in cotton manufacturing. *New York Public Library Picture Collection*

Cotton Gin

From the time he was young, Eli Whitney could be seen demonstrating the kind of curiosity that is typical of inventors. He spent much of his childhood free time in his father's metalworking shop taking apart pocket watches and clocks and putting them back together again to see how they worked.

By fourteen, he had opened a nail making business—his nails were made on a machine that he had designed and built himself—and then a ladies hatpin making shop (the only one in the country for some time). Ultimately, he went to Yale University and then, having compiled some debts, in 1792 took a teaching job to pay them off. This took him to a plantation in Savannah, Georgia, where he overheard many of the cotton farmers' tales of woe, about the work being so tedious and how under the best conditions one could manage to clean—that is, remove the seeds—only one pound of the crop a day. They were in a desperate plight.

He worked on a design that would ease this task; he eventually came up with a machine that made it possible to clean 50 pounds of cotton a day. The design of the machine was simple, but effective: it consisted of a cylinder with wire teeth. The raw cotton from the field was fed through the cylinder and as it spun around the teeth passed through small slits in a piece of wood, pulling the fibers of the cotton all the way through but leaving the seeds behind.

The influence of the cotton gin on the commercial production of cotton was staggering. In 1793, about 180,000 pounds of cotton were harvested in the United States. Only two years later, the output grew to more than 6 million pounds. By 1810, 93 million pounds of cotton were harvested a year.

The cotton gin played a major role in history because it not only helped farmers clean more cotton faster, but it also indirectly helped the Deep South revive a lagging industry and compete with more profitable cash crops such as tobacco and indigo. It's been said by many that the cotton gin also started the Industrial Revolution in America because of its immediate impact on industry. Once the steam engine was adapted to drive the gin, the process became totally automated and a whole new business, a business that changed the face of the country, was born.

Cotton quickly rivaled the cash crops because it took very little effort to grow. It needed very little water and could grow in many different types of soil. Although it was abundant before the gin, after the gin's invention farmers were growing it more and more and planting it in fields that were for years deemed useless. Crop rotation—leaving sections of fields to lie fallow for a year or more so that the soil could regenerate itself with nutrients—wasn't needed because cotton soon filled otherwise barren fields. Farmers were beginning to make money on cotton.

Whitney was known for his mechanical abilities and he was said to have been able to "fix anything." From the time he started working on the machine to the time it was finished was an astounding 10 days. Historians believe that in those 10 days the invention of the cotton gin changed the face of the South and the U.S. economy.

What seemed like a short amount of time (10 days) was actually the accumulated time of an entire boyhood of tinkering.

Whitney, a careful observer, studied the hand movements of people separating the seeds. One hand held the seed while the other teased out the strands of lint. His machine was designed to duplicate this. To simulate a hand holding a seed, he made a sieve of wires stretched lengthwise. To do the work of the fingers, he had a drum rotate past the sieve, nearly touching it. Fine, hook-shaped wires projected from the drum and caught all the lint from the seed. The restraining wires of the sieve held the seeds back while the lint was pulled away. A rotating brush, which revolved four times as fast as the drum, cleaned the lint off the hooks.

Whitney's machine became known as the cotton gin and never became more complicated. The invention, which started as a time-saving and problem-solving device, grew to become one of the most important inventions in the history of U.S. economics.

86

Windmills in Holland. *Photofest*

Windmill

The invention of the windmill represented a remarkable achievement for humanity. Instead of struggling to overcome mother nature, people learned to work with her and harness her power.

The earliest known use of wind power was the sail boat, and what was learned had a significant impact on sail-type windmills. Ancient sailors understood lift and drag and used it every day. It wasn't until later that people applied this knowledge to windmills.

The first windmills were invented to automate grain grinding and water pumping. Grain grinding and water carrying were traditionally very labor intensive so the need to make them easier was apparent.

The first design for a windmill was in Persia around A.D. 500 to 900. Scholars believe it was designed to pump water, although they do not know exactly how it worked because there are no existing plans or drawings. According to oral accounts, the windmill was designed with vertical sails made with bundles of reeds or wood that were attached to a central, vertical shaft by horizontal struts.

The first documented windmill ground grain. In this design, the grinding stone was attached to the same vertical shaft. The machinery was housed in a building so the wind could not impair its operation.

The first documented use of the vertical-axis windmill in China was in 1219, although some believe it was far earlier—before Persia in fact—but it cannot be proven. These windmills were also used to grind grain and pump water.

By the time windmills appeared in western Europe in 1300, they were of a horizontal-axis design. Although the reason is not precisely known, some guess that the advent of the water wheel (which was designed with the horizontal configuration) had an effect on the design. Another reason for the horizontal shift is that it is simply more efficient. Vertically designed windmills lose power by having to protect the trailing side of the sail from the oncoming wind.

As early as 1390, the Dutch set out to refine the tower mill design. To achieve maximum results, they attached the post mill to the top of a tower that was several stories high. Separate floors inside were devoted to different jobs, such as storing grain, removing chaff (the seeds from the stalk), and grinding grain. The bottom floor was devoted to the windsmith and his family.

The tower and post mill had to be manually adjusted to face directly into the wind for maximum efficiency and to remain structurally sound. These were the chief responsibilities of the windsmith.

Later, a primary improvement of the European mills was their use of sails that generated aerodynamic lift. This was so important because it improved rotor efficiency (the rotor turned faster with less effort) compared with the Persian mills. The bottom line is that it improved the grinding and pumping action.

Perfecting the windmill sail, however, took nearly five hundred years. Efficiency was the constant goal and by the time the process was completed, windmill sails had all the features needed for the performance of modern wind turbine machines. Some of these features included leading and trailing edges of blades (like airplane propellers and wings) and the correct placement of blades on a rotor.

Windmills were so important at one time in Europe that they served as the "electric motor" of Europe before electric motors were even invented—and before the Industrial Revolution began. Their applications were so diverse, ranging from grain grinding, water pumping, sawing timber, and the processing of commodities, that they were indispensable for hundreds of years.

By the end of the nineteenth century, large tower designs of windmills dropped off with the increased demand for steam engines. In the western part of the United States, smaller windmills were popping up and their efficiency was fined-tuned for maximum results. The first mills had four paddle-like wooden blades. They were followed by mills with thin wooden slats nailed to wooden rims. Many of these windmills had tails that were used to orient themselves into the wind.

The most important improvement of the American fan-type windmill was the development of steel blades in 1870. These blades were lighter and could be more flexible when bent into different shapes.

Between 1850 and 1970, over six million small (one horsepower or less), mechanical-output wind machines were installed in the United States alone. In the late nineteenth century, the successful American multiblade windmill design was used in the first large windmill to generate electricity.

In modern times, many advances were made in windmills for another, equally important purpose: to generate electricity. In this effort, modern development of wind systems are inspired by the design of airplane propellers and wings.

Submarine 87

One doesn't need to ponder the submarine long to gauge its effect on humanity. In World War II, it came into its own as a machine of war, and today nuclear-powered submarines armed with nuclear missiles are in the forefront of modern warfare.

It is an unlikely but a true story that the most important person ever involved with the submarine was a teacher named John P. Holland. Developing it in his spare time, he put the submarine on track to become one of the deadliest weapons of war.

Holland was born in Liscannor, Ireland, in 1840. He was educated at Limerick Christian Brothers School, took his vows to become a Christian brother in 1858, and taught in a number of different places. As the years passed by, however, he found it harder and harder to focus on his vocation. Growing up as a son of a coastguardsman, he had seawater in his veins and always longed to go to sea, a desire that was stymied by his poor eyesight. Eventually released from his vows, Holland and his family traveled to America in 1872.

Holland had been interested in submarines since he was a boy. While he had no formal education, he was self-taught in engineering and drafting and showed a brilliant aptitude for these things.

Part of Holland's interest in the submarine was triggered by the constant conflict between England and Ireland, about which Holland felt deeply. The British had a formidable navy, which he knew Ireland could not hope to defeat in a head-on naval war, but he thought that perhaps the submarine could do a lot of damage without being detected.

In 1874–1875, Holland tried to get the U.S. Navy interested in the submarine, but his efforts were impeded by the fact that the submarine was not a new idea. Indeed, the *Hulney,* a Confederate submarine, sunk a Union warship during the Civil War. (Interestingly, the *Hulney,* which had sustained serious damage during the confrontation, was also sunk.) And a hundred years before that, a submarine had been invented by David Bushnell, who tried to sink a ship during the Revolutionary War. In sum, the U.S. Navy thought the idea preposterous, in part because Holland was not a sailor, which, of course, was equally preposterous, as if only a sailor could invent a seagoing craft.

But an Irish rebel group, the Fenians, was interested. Though it had been dealt a serious blow in the war against England because of its defeat in Canada in

Patent drawing, 1902, by John P. Holland. *U.S. Patent Office*

1866, a number of the members had regrouped by the time Holland presented the idea to them. He impressed them enough that they took 60,000 pounds from their "skirmishing fund" to invest in the project. When he finished building it, Holland and the Fenians assembled on the banks of the Passaic River in New Jersey to watch the launching of the 14-foot craft.

The result was disastrous: Unable to float for very long, the submarine quickly filled with water and sank to the bottom. But it was raised and after an examination it was discovered that one of the workmen had failed to install a pair of screws that had left an opening through which water poured.

The submarine was drained, the screws reinstalled, and Holland himself took it out. It floated, dived, and—much to his relief, no doubt—resurfaced.

While fine-tuning the craft, he began devising a plan for mounting an attack against the British. Holland was well aware of the fire power of the British ships, which made approaching them without being seen vital. His plan was a classic Trojan horse: a harmless, ordinary ship would get as close as possible to the British ships before the submarine was launched through a trap door in the ship.

But Holland never got a chance to put his plan into action. In 1883, several of the Fenian members began to break away. This deterioration was highlighted by a group of them taking the submarine, which was anchored in New Jersey, and hauling it up to New Haven, Connecticut, where they tried launching it. Unsuccessful, they abandoned the craft at a nearby brass factory. When Holland heard of this joy ride, he was incensed, and the great scheme was abandoned. Holland and the Fenians never communicated again.

Holland was truly ahead of his time. His theory about the submarine was that the best possible shape would be that of a cigar. But the soundness of this idea did not emerge until the 1950s, long after Holland was gone.

Holland never made any money from his invention, and as time when by he started to have second thoughts about the havoc that submarines could wreak. Indeed, the validity of his concerns was more than borne out in World War II, when "wolfpacks" of German submarines torpedoed hundreds of ships (many of them carrying civilians), sending thousands of their human occupants to a watery grave.

88

Paint protects and preserves. *Author*

Paint

Paint is another one of those inventions that's taken for granted. It's just there, all around us, doing a fantastic job—when you think about it—to extend the life of the myriad items it's on, from cars, to machinery, to walls and floors, and much, much more.

Paint has been protecting things for a long time. Indeed, the earliest known use of surface coatings goes as far back as 2000 B.C. Early Chinese and Egyptian artisans used mixtures of drying oils, resins, and pigments for pictures and inscriptions in their tombs and temples. Surprisingly, these paints resemble—in makeup and appearance—more fundamental types in use today.

As the world population began to expand, people began to travel, trade, and go to war on an organized scale, the need and desire for decorative coatings grew. The ancients used paint on their ships, utensils, musical instruments, weapons, and palaces in an ever-growing variety of pigments and binders. White pigments came from white lead and natural white materials such as clay, gypsum, and whiting. Black pigments came from charcoal, lampblack, boneblack, natural graphite, and powdered coal. Yellow pigments came from ochers, gold powders, and lithage. Reds came from iron oxides, red lead, cinnabar, and natural red dyes. Blues, such as Egyptian blue, came from lapis lazuli, copper carbonate, and indigo.

Greens came from terre verte, malachite, verdigris, and natural dyes. Their binding media included gum arabic, gelatins, beeswax, pitch shellac, animal fats, drying oils, and saps from various trees.

The amount of paint made was not much by modern standards. A generally low standard of living, scarcity of raw materials, and slow processing of paint by hand resulted in slow growth in the use of paint. However, humanity's inventiveness led to developing better manufacturing methods. In the 1200s, a monk named Prebyster described the making of a varnish based completely on nonvolatile ingredients, chiefly drying oils.

About 1500 B.C., the first modern varnishes were made by "running" resin with sandarac or linseed oil. These varnishes were mainly used to decorate and protect crossbows and other weapons. During the next three hundred years, the most popular resin used for both protection and decoration was amber, either alone or combined with linseed oil. The scarcity of amber led to a search for substitutes, and it was replaced almost completely with fossil and semifossil gums such as gum arobon and gum elastic.

In the twentieth century, the paint industry underwent dramatic advances. For many years, paint for buildings was characterized as oil-based, meaning that the binder in the paint was some sort of oil, such as alkyd or linseed oil. To thin and clean up the paint, one used either turpentine or, more frequently, benzene. It was a difficult material to work with, one of the reasons being that it took 24 hours or more to dry.

Then, latex paints, which could be thinned and cleaned up with water, came along in the 1960s, and oil-based paints started to, as it were, fade. Still, old-time painters will tell you that in terms of longevity and quality of finish, oil-based paint was much better than latex-based paint.

But then latex got better, and oil got worse, the reason being that environmental watchdogs discovered that the fumes given off by oil-based paint, the volatile organic compounds, were making holes in the ozone layer. The U.S. government became involved and forced manufacturers to reduce the amount of these compounds to certain levels. As a result, most oil-based paints today are mere shadows of their former selves, while latex has been improved to the point where it is generally superior.

It was also discovered that some oil-based paints contained lead, which had adverse effects on people, particularly children, when ingested. In 1978, the government passed legislation that barred lead from being used in paint.

Paint has become a boon to do-it-yourselfers worldwide. It is very easy to apply—even for someone who has never used it before—and this has been another building block, as it were, in the success of home improvement chains, such as Home Depot and Lowe's, who essentially cater to do-it-yourselfers.

89

The "breaker" is a vital safety device. *Author*

Circuit Breaker

When Thomas Alva Edison, Nikola Tesla, George Westinghouse, and other invention all-stars created the electrical system, they were, of course, working with a powerful force, one that could be deadly in two ways: either by shock or by being the culprit if a fire broke out. As such, it became necessary to create a variety of safety devices that would protect people and property from electrical malfunction.

A variety of these devices were made, and none was more important than the "fuse," which eventually transmogrified into the "circuit breaker." It had one basic role: to keep buildings and their occupants safe. To understand how it works, one must first understand how electricity works.

In a house, there are, first of all, "circuits," a series of wires inside walls and ceilings that provide electricity to the various electrical devices in the home. The average home has three kinds of circuits: general purpose, small appliance, and individual appliance. General purpose circuits, also known as "lighting circuits," power lights, television sets, and other things that don't require a great deal of power. Small appliance circuits serve small appliances, such as deep fryers, blenders, and the like, while individual appliance circuits power things that demand a lot of electricity to function, such as stoves and driers.

Electrically, in describing these three circuits, one speaks of "voltage," which is the amount of push electricity gets through the wires, and "amperage," which is the all-important amount of electricity.

The idea is that the circuits should be properly sized for the devices that they are going to serve. For example, if the wire is going to serve a 20-amp appliance, it means that the wire serving it must have a gauge that is thick enough to carry that much amperage without heating up, because that's what will happen if too much current is piped, as it were, through a wire that's too small.

This is where the fuse/circuit breaker comes in. A circuit breaker is capable of sensing "overloads" on the wire; when it senses one, it shuts down the electricity flowing through the wire by "tripping." It does this because at the heart of its construction is a bimetallic strip, two springs, and two control points. When a circuit overload occurs, the strip bends and pulls away from the contact points, which have electricity flowing through them, thus shutting off the electrical flow. To reactivate the circuit, one merely has to flip the switch on the circuit breaker. (Actually, the best way to turn a "breaker" on is flip it to "Off" and then back to "On.") If there is a problem, the circuit breaker will shut down the flow of electricity again. In the case of a fuse, it "blows," meaning that the little strip of metal, through which the electric flows burns up, sacrificing itself to break the electrical flow.

The first circuit breaker appeared in 1829 and its principle of operation was based on the electric relay, which was invented by Joseph Henry, an American physicist. Then, as the twentieth century approached, circuit breakers used springs and electromagnets. If the current became too high—that is, overloaded the circuit—the relay would be tripped, thus breaking and shutting down the electrical connection.

The fuse was invented by English physicist James Joule. In 1840, he discovered a formula that revealed the amount of heat that an electric current created. Based on this, he was able to determine how thin or thick a fuse must be before melting. His formula said that heat was proportional to the square of the intensity of the current multiplied by the resistance (the opposition by a material to the flow of electric current) of the circuit.

To this day, some people will try to defeat fuses—we don't know of any way to defeat breakers—by putting copper pennies under the fuses, so that if there is an overload the lights will stay on.

90

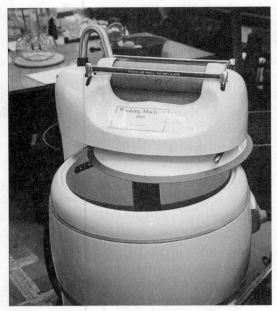

Washing machine, circa 1948. *Author*

Washing Machine

For as long as people have been wearing clothes, they have searched for a way to clean them. Prior to 1797, it was about hauling laundry down to the nearest water source, laying the clothes on a rock, and literally beating the dirt out of them with the best rock one could find.

Improvement came in 1797 with the invention of the washboard. The first actual washing machine was manual and imitated the action of hand washing on a washboard. It operated by using a lever to move one surface (curved in shape) over another of the same shape, thus removing the dirt. This type of machine was first patented in the United States in 1846, and was used until 1927. In 1851, James King patented the first washing machine that used a drum, and in 1858 Hamilton Smith patented the rotary washing machine.

Even though washing clothes was always seen as a woman's task, William Blackstone had a very hard time watching his wife do all that backbreaking work. In 1874, he improved the washing machine (as a birthday present for his wife) by building a wooden tub with a handle to turn the internal gears. This mechanism grabbed the clothing and moved it around in the water, releasing the dirt through the motion.

This was a fabulous breakthrough, although perhaps not so extraordinary if one had been a sailor. For centuries, and out of necessity, sailors at sea would put their clothing into nets secured with heavy ropes. These bundles, thrown overboard and dragged behind the ship, were cleaned by forcing water through the clothes.

It is important to keep in mind that up through 1907 all types of washing machines were run manually. It was not until 1908 that the first electric-powered washing machine, the Thor, was introduced by Alva J. Fisher, an inventor for the Hurley Machine Company. This machine certainly took a lot of the labor out of the task of laundering; however, the motor that rotated the tub was not protected from leaks from under the machine, so it experienced short circuiting and gave jolting shocks. By 1911, it was possible to purchase domestically made oscillating cylinder machines with sheet metal tubs, which were mounted on angle iron frames with perforated metal or wooden slat cylinders inside.

As the invention progressed, so did the challenges that its improvers faced. A better motor and starting mechanism was needed, one with enough power to start the machine, but not so much power that it would burn out or overheat. Shocking was also a problem, so the mechanics had to somehow be enclosed, and once enclosed a type of cooling device or fan was needed to reduce the risk of overheating. There were also improvements to be made on the tub itself. The wooden and cast iron tubs were replaced by a lighter metal one. Beginning in the 1920s, white-enameled sheet metal replaced the copper tub and angle iron legs. By the 1940s, this became the preferred choice because it was seen as more sanitary and easier to keep clean.

Even with all these vast improvements, the washing machine was still not common enough for a person or family to have one at home. It was not until 1936 that people, despite the Great Depression, started to bring these machines into their homes. Prior to this, women, where able, went to the local washateria, where a gleaming row of the new machines could be found, along with small boxes of detergent and even soda and candy.

By the early 1950s, American manufacturers were building machines that not only washed the clothes, but spun them dry as well. This replaced the need for a "wringer," a device that extracted water from clothing by squeezing it between two rollers.

As if this wasn't enough, in 1957 General Electric introduced a washing machine with push buttons to control wash temperature, rinse temperature, agitator speed, and spin speed. The washing machine had come into its own. From manual, to gas, to electric, and with added push-button features, the washing machine became an appliance that no home or woman wanted to be without. Presently, the washing machine comes in a variety of colors with a myriad of options that previous generations of users would never have dreamed possible.

For women, the ability to have the washing machine in the home was both a blessing and a curse. It is true that it was and is certainly more convenient, time saving, and easier than dragging one's laundry to the closest Laundromat to spend the day. But that day was not simply a chore, it was also a social event. The Laundromat was a place to talk, share, catch up on things, and have a snack while the washing machine took care of most of the physical labor of laundering. Having a machine in the home took away this social aspect.

Threshing Machine 91

When wheat was harvested, it was "threshed," meaning the stalks were separated from grain. Then, it was hit with a "flail," which forced the grain to pull apart from the chaff; this was known as "winnowing." In some places, harvest wheat was spread on the ground and a heavy, animal-pulled sled was drawn over it. After threshing, the grain was winnowed. All this could take up to a couple of months.

In 1830, about 250 to 300 hours of labor were required to produce 100 bushels (5 acres) of wheat. This was done with walking plows, brush harrow, hand broadcast of seed, sickle, and flail. Not until 1834 was the McCormick reaper patented; in that same year, John Lane undertook manufacturing of plows faced with steel saw blades. John Deere and Leonard Andrus started manufacturing steel plows in 1837, the same year that a practical threshing machine was patented.

In 1786, Andrew Meikle (who "descended from a line of ingenious mechanics," according to his tombstone) invented a threshing machine in Scotland. His father had invented a winnowing machine in 1710 that was not well received, as mechanical contraptions at that time were regarded with suspicion. The son's thresher proved more successful. He worked as a millwright at Houston Mill on the family estate of John Rennie at Phantassie, East Lothian. Rennie collaborated with Meikle to get his machines installed in other mills.

In the beginning, not everyone owned a threshing machine. Larger farmers did, but smaller farmers depended on traveling thresher men to do the job. Regardless of who owned the machinery, the threshing process required a huge amount of labor. Each member of the threshing crew had a specific job to accomplish when the harvesting process began.

The steam engineer set up the thresher in a location near a grain field or wherever the farmer wanted the straw blown. The steam engine was belted up to it; all the prethreshing work was done on the threshing machine. While the machine was starting up, a team of workers called "bundle haulers" went out into the field, loaded shocks onto a horse-drawn wagon, and then drove the filled wagon to the threshing machine. Men stood on top of the wagon and pitched grain bundles down into the threshing machine's bundle feeder. The bundle feeder had a conveying chain that transported the grain bundles into the threshing machine cylin-

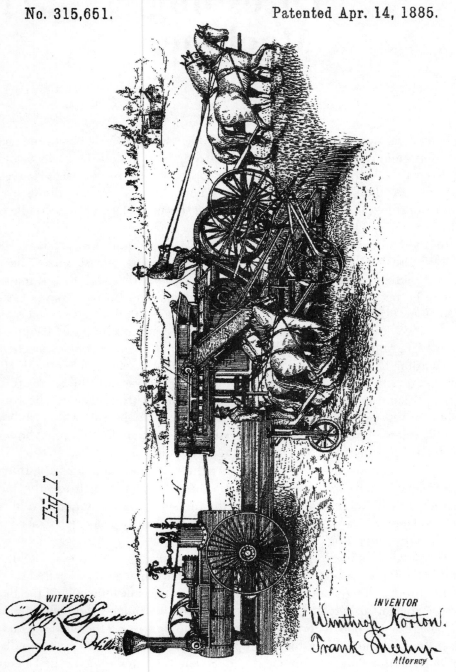

Patent drawing, 1885, by Winthrop Norton. *U.S. Patent Office*

der, where most of the grain was separated from the stalks and fell to the bottom of the threshing machine. Chaff and dust were removed by a fan as the grain floated down, and an elevator on the thresher moved the loose grain into a grain wagon parked nearby or into individual bags, whichever the farmer preferred. The straw was continually battered as it progressed through the cylinder, ensuring that all the grain was stripped off the stalks. After the straw passed over the straw walkers, it was deposited in a fan housing at the back of the thresher that propelled it through a blower into the straw stack.

This process was repeated continuously until all of the farmer's grain was threshed. Despite the great labor required to use a threshing machine, it enormously improved efficiency and capacity of threshing over all previous methods. In the 1840s, growing use of factory-made agricultural machines increased farmers' need for cash, thereby creating a huge incentive for them to undertake commercial farming.

92

This invention saves lives and property. *Author*

Fire Extinguisher

Fire is a chemical combustion reaction between oxygen in the air and some sort of fuel, like wood or gasoline, that's been heated up to ignition temperature.

A wood fire is a reaction among oxygen, wood, and extreme heat: A focused light, friction, or something already burning heats the wood to a very high temperature (500° F/260° C degrees or more). The heat decomposes some of the cellulose of which the wood is mostly made. The decomposed material releases volatile gases like hydrogen, carbon, and oxygen. When the combined gas gets hot enough to break apart the compound's molecules, the atoms recombine with the oxygen to form water, carbon dioxide, and other products. The gases rising up into the air make up the flame. Carbon atoms rising into the flame emit light as they heat, and that flaming heat keeps the fuel at the ignition temperature—it keeps burning as long as there is fuel and oxygen.

To extinguish a fire, at least one of these three elements has to be removed. The most obvious way to remove heat from a fire is to dump water on it, cooling the fuel to below the ignition point and interrupting the combustion cycle. To remove oxygen, the fire has to be smothered so it is not exposed to air: a heavy blanket or nonflammable material like sand or baking soda dumped on it will do. To remove fuel is more difficult. With house fires, the house itself can act as fuel; the fuel is "removed" after the fire has burned it all.

Water is the most familiar effective extinguishing material, but if it is used in the wrong situation, it can actually be dangerous. It can be poured on burning wood, cardboard, or paper, but if used on electrical fires the water may conduct the current and cause electrocution. If it is used on burning inflammable liquids, the water will simply spread out over the fire, probably making the fire worse.

Fire fighting existed in ancient Rome, which had actual fire brigades with seven thousand paid firefighters that responded to and fought fires, patrolled the streets, and imposed corporal punishment on anyone violating fire-prevention codes. Ctesibius of Alexandria invented the first known "fire pump" around 200 B.C., but fire pumps were reinvented about A.D. 1500. In 1666, London had only 2-quart hand syringes and a similar, slightly larger syringe to fight the great fire that burned for four days. The rest of Europe and the American colonies had similar equipment. The London fire stimulated the development of a two-person-operated piston pump on wheels.

New York City appointed fire inspectors with the authority to impose fines for fire-code violations in 1648. Boston imported the first fire engine in 1659, while Thomas Lote of New York built an American one in 1743. Leather was used for hose and couplings until fabric and rubber-treated hose came into use in 1870, when the "aerial ladder wagon" appeared. The "hose elevator" followed.

A "fire extinguisher" was patented by Thomas J. Martin in 1872. In late 1935, back in the United States from Vienna where he had studied and worked, Percy Lavon Julian left academic life and entered the corporate world as chief chemist and director of the soy products division of the Glidden Company. He was the first black scientist hired at such a level and an inspiration to others. Glidden manufactured paints and varnish; the company wanted Julian to develop soy-based compounds for it. Julian came up with an "aerofoam" extinguisher that put out gas and oil fires. Used by the U.S. Navy, the extinguisher saved many sailors' lives during World War II. Ironically, in 1950, soon after he bought a house in Oak Park (Chicago) for his family, an arsonist set it on fire.

Nowadays, fire extinguishers use compressed carbon dioxide to smother flames: they suffocate the fire by evacuating the oxygen within its immediate area. The typical metal cylinder with a short hose attached is a "soda-and-acid" extinguisher; inside, above a solution of soda and water, is a container of acid. When the extinguisher is turned upside down, the acid mixes and reacts with the soda to generate carbon dioxide. Gas pressure forces the solution out of the hose. A "foam" extinguisher's cylinder contains water, sodium bicarbonate, and something such as licorice powder to strengthen the foam, plus an inner container of powdered aluminum sulfate. Together, they create a foam of carbon dioxide bubbles. A "carbon dioxide" extinguisher consists of a tank of liquid carbon dioxide under pressure. When released, the carbon dioxide forms flakes that vaporize and blanket the fire.

Today's fire extinguishers are sturdy metal cylinders filled with water or a smothering material and all operate on a PASS system (Pull pin, Aim low, Point nozzle at base of fire, Squeeze handle, and Sweep from side to side across fire area). They are rated by the type of fire they can put out: "Class A" extinguishers can put out wood, plastic, or paper fires; "Class B" can put out burning liquids like gasoline or grease; "Class C" can put out electrical fires; and "Class D," designed to put out burning metal, are rare. Extinguishers marked "ABC" can put out all types of fires other than those covered by the Class D extinguisher.

One material used to extinguish fires is pure carbon dioxide, which is kept in pressurized liquid form in the cylinder. When the container is opened, the carbon dioxide expands to form a gas in the atmosphere. Carbon dioxide is heavier than oxygen, so it displaces the oxygen surrounding the burning fuel. This sort of fire extinguisher is common in restaurants because it won't contaminate the cooking equipment or food.

The most popular extinguisher material is dry chemical foam or powder, typically made of sodium bicarbonate (normal baking soda), potassium bicarbonate (nearly identical to baking soda), or monoammonium phosphate. Baking soda starts to decompose at only 158 degrees F (70° C). When it decomposes, it releases carbon dioxide, which, along with the insulation of the foam, works to smother the fire.

Most fire extinguishers contain a fairly small amount of fire-suppressant material that gets used up in a few seconds. That's why extinguishers are effective only on relatively small, contained fires. To put out a larger fire, you need much bigger equipment like a fire engine and people who know how to use it. But to keep on hand for putting out small, dangerous flames, a fire extinguisher is essential.

Mid-twentieth-century American refrigerator. *Photofest*

Refrigerator

In many climates, nature provides temperatures capable of preserving many perishables at just above or below the freezing mark. But some climates, especially where early civilization flourished, had little or no winter at all. What these civilizations did, particularly the Greeks and Romans, was to use nature as best they could.

Wealthy families from the Mediterranean area, for example, built cellars as deep as possible, then lined them with straw and wood to provide a layer of insulation. Ice and snow were then brought down from the surrounding mountains and placed in these cellars, creating "ice cellars" that could keep things cool for months at a time. This way of cooling preservation is still used in some undeveloped countries; indeed, it is the primary means of refrigeration for these countries.

Naturally, where there's a need there's money to be made, and the sale of ice flourished as a business for quite some time. In the Americas, for example, Yankee traders built thick, double-hulled ships that carried ice from Canada and Maine to the southern states, the Caribbean islands, and South America. During the American Civil War, the Confederacy, for the most part, had to do without refrigeration for the duration of the war.

While much of the world kept cool with naturally harvested ice and hard-packed snow, American Indians and the ever-enterprising Egyptians found an answer in physics. These two cultures, independent of each other, discovered that if a plate of water was left out over night in the cooling tropical air, the rapid evaporation rate would leave ice in the plate, even if the surrounding air did not dip below the freezing mark. This rather amazing way to produce ice gave rise, really, to modern refrigeration. The key question was how to harness the cooling power of rapidly expanding gasses.

Some other techniques were tried, such as introducing potassium or sodium nitrate to water to reduce the temperature. This became popular for a while in France in the 1600s and was used for cooling wine among other things.

It wasn't until 1748, however, that progress in refrigeration occurred. In this year, William Cullen, a respected Scottish physician, chemist, and professor of medicine, successfully demonstrated the cooling properties of ethyl ether when allowed to boil off into a partial vacuum. While Cullen never applied his findings to any practical purpose, he did inspire many others to try to achieve something. Oliver Evans, an American inventor who helped develop the steam engine, designed an artificial, mechanical refrigeration machine.

John Gorrie, a Florida physician, toiled to apply the refrigeration concept to cooling off hospital rooms during a malaria outbreak in his hometown of Apalachicola. Working from Evans's design, he succeeded in compressing gas, cooling it by sending it through radiating coils, and then expanding it again to make it even colder. He patented his device in 1851, thus making it the first U.S. patent for mechanical refrigeration. Gorrie gave up his medical practice in order to find investors to build a factory for his machines, but to no avail.

In the nineteenth century, other inventors came on the scene with similar gas-compressing devices, including Jacob Perkins, who had a working model shortly before Gorrie's.

Important advances started to be made, including the development of commercial refrigeration units, which allowed for the transportation and storage of perishable food items over long distances. Phillip Danforth Armour, an American entrepreneur, is one whose company flourished as a result of its refrigeration systems. Establishing cold-storage warehouses on the East Coast, he was able to ship his meat products as far as Europe, making Chicago the capital of the meat-packing industry.

And all this came at an opportune moment, because it was soon discovered that many of the ice industry's water sources were beginning to suffer from the effects of pollution, namely soot and untreated sewage. Regardless, the "icebox," which required blocks of ice to keep its contents cold, remained in common use until the home refrigeration unit grew to supplant it in the 1940s and 1950s.

In 1859, Frenchman Ferdinand Carré improved the existing refrigeration systems by using rapidly expanding ammonia, instead of air, which contains water vapor. Ammonia remains a gas at a much lower temperature than water vapor, so it can absorb more heat. The drawback was that if it—and other types of vapor-compressor systems—leaked, it was toxic. After some tragic events, it became quite clear that something safer and less toxic was needed.

Industry engineers collectively worked to find an agent that was much safer to use and eventually came up with "Freon," an artificial chemical that added chlorine and fluorine atoms to methane molecules, rendering the refrigerant virtually harmless except at massive doses. In this way, the consumer had no reason to fear refrigeration units in the home, which paved the way for mass production.

94

Mid-twentieth-century American oven and stove. *Author*

Oven

The oven is a simple device: an enclosed chamber designed to contain a dry, uniform heat, which can be used for cooking food or processing things into a harder state, such as iron ore into iron or clay into ceramics.

Soon after the discovery of fire, it was realized that different rocks had properties that could retain or transmit heat. By prehistoric times, it is believed that bread had become an important part of the human diet. Cereal grains or their seeds are believed to have been roasted on an open fire until it was discovered that they were a lot easier to digest when mixed in water and then heated into a gruel. It was also learned that heated stones retained their heat, so that pouring this gruel onto the hot, flat stones around the fire evaporated the water and produced the first "pancake"—or "flatbread," if you will

Eventually, these early peoples learned that gruel or batter set aside for later cooking fermented or basically spoiled due to the presence of yeasts. While this was not always a good thing, it led to an important discovery, namely that thick batter became shapable dough that could be cooked into the "leavened bread" we know today.

The Egyptians made the earliest and most profound use of leavened bread, however. By about 2600 B.C., they had harnessed yeast as a leavening agent by taking portions of "sour dough" and mixing it with fresh dough, thereby "infecting" the fresh dough under relatively controlled circumstances. Baking techniques in general flourished under this ancient civilization, resulting in as many as fifty different breads, flavored with as many types of seeds, like sesame and poppy, that could be found.

So, not surprisingly, it is the Egyptians who are credited with inventing the first ovens. We look backward slightly to remember the fact that certain kinds of clay, with Nile clay being one of the most notable, transformed into solid ceramics in the heat of a fire. By forming the clay into a large cylinder that tapered at the top and placing a shelflike partition about halfway down the body, we have the first cooking ovens. The bottom portion was the "firebox" and the top part was the oven or "baking chamber" where pieces of dough were placed on the partition to bake. This simple structure is the very foundation of the oven, which remains to this day. Whether it is a mass-produced, high-tech electric or gas model or the terra-cotta variety, the modern-day oven differs very little from its ancient predecessor.

Variations on this structure can be found throughout the ancient world with varying degrees of sophistication. It was common for every family or village to have some sort of oven and as towns grew, like Jerusalem, entire areas would be set aside for public ovens.

Rome is a good example of this urban transition that ultimately turned into the baking industry. Initially, bread baking remained a domestic task with the preparation and cooking techniques as rustic as anywhere else. But by around the second century B.C., Pliny the Elder writes of professional bakers who began to produce breads for the rich, who did not need to toil away for hours in the process. Usually former slaves from throughout the empire, these bakers brought techniques from far and wide to the task of supplying wealthy Romans with their staple food, ultimately forming guilds that combined the millers with the bakers. Before long, these bakers held a status not unlike a civil servant, with specific guidelines, recipes, and processes all regulated by the government.

But aside from the advancements in the industry itself, the Romans never made any truly major breakthroughs in baking or the oven, though they did create the first known mechanical mixer to knead dough. Huge paddles were driven by a horse or donkey to meld the flour, water, and leavening in a large stone basin.

The Romans generally used what is called the "beehive" oven, though they also cooked bread on spits over open fires and in earthenware vessels exposed to open fire. The beehive oven is one of the most common types to be found. Not all ovens were the same, of course, but a basic construction is as follows: A shallow pit in the ground is filled with flat stones, roughly 2 feet by 3 feet. The spaces between the stones are filled in with clay to make a nice flat bed. The walls of the oven are then built up into a dome shape, not unlike an igloo, using a wide variety of "corbeling" techniques and materials, everything from terra-cotta, to stones, to mounds of earth built on a frame. A small space is left for the door, which is covered by a slab of rock. A fire is built inside to dry out and fire any clay used in the oven. More clay can be used to patch any holes in the dome at this time and still more earth is piled on to further insulate the baking chamber.

When the oven is used, it is fired to the desired temperature for baking, which until the invention of the thermometer could only be determined by skilled craftsmen. Ash, embers, and any remaining burning wood or charcoal are removed, and the oven is ready to bake.

The oven itself, while having its historical moments, is such a universal concept, almost as basic as the discovery of fire or the invention of the wheel, it is an injustice to narrow its conception down to any one place, culture, or time. The basic oven appears all over the world and was by no means restricted to just the baking of bread. As previously discussed, the firing of clay made the original ovens possible. The more strides made in the development of the oven, the more strides in the making of ceramics and bricks, which in turn led to the development of the brick oven. As smelting became possible with the oven's technology, so, too, did cast iron pots and ovens, from the tiny and efficient "Dutch" oven, to the cast iron ranges, and finally to the massive blast furnaces used to make steel.

Bicycle 95

Although bicycles are usually associated with recreation, the earliest model came about as a way of solving a problem. In 1817, Baron von Drais needed a way to get around his royal gardens faster than walking without tiring. His answer: "the walking machine."

The person straddled the frame, at the front of which was a steerable wheel and at the back there was a second wheel that was aligned with the first. There were no pedals. The bicycle rider propelled himself by pushing his feet along the ground. Although the expensive machine enjoyed some popularity among the rich, it was less than practical because it could only be used on a smooth, flat surface (such as a well-maintained dirt path).

A later model appeared in 1865 that had pedals on the front, as well as a larger wheel. It was named the "velocipede" or "boneshaker" because of the sometimes violent, always uncomfortable ride the user had to endure. This was because it was made entirely of wood and then, later, with metal tires, and the marriage of these on cobblestone led to a teeth-jarring ride.

Around 1870, the first all-metal machines appeared, with some other important changes. One change was that the larger front tire got even larger still—in fact, it began to be designed according to how far the user could stretch his or her legs—because it was realized that the larger the wheel, the further one could travel per pedal stroke. The other, more important change was the use of solid rubber tires.

Also, designs considered women with the introduction of an adult tricycle, which allowed them to ride despite their long skirts. The tricycle also seemed more appropriate, dignity wise, for the clergy, physicians, and others of high societal standing to use.

Other improvements included rack and pinion steering, hand brakes, and the differential (used to redirect gears and transfer power from one axle to another). So important were some of these changes that they are still in use in bicycles—and cars—today!

Still, problems of a comfortable, practical wheel remained. Although the rubber used in bicycle tires was helpful, it couldn't stand up to the weather. In cold weather it dried and became brittle and in warm weather it became too soft and tacky. A method of treating the rubber so it remained consistent was needed.

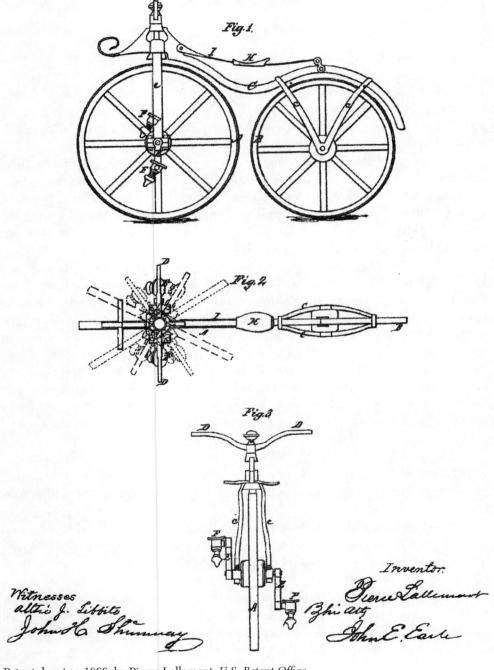

Patent drawing, 1866, by Pierre Lallement. *U.S. Patent Office*

During this time, Charles Goodyear, the son of a New Haven, Connecticut, storekeeper, was experimenting with rubber to find a solution.

As luck would have it—like a scene straight out of a science-fiction movie—one day in 1839 he stumbled across the solution by accidentally spilling sulfur and rubber on a hot stove in his workshop. The results of the accident were staggering. He was amazed to find that it had not lost its stickiness. When he exposed the sulfur-contaminated rubber to hot and cold conditions, he found that the sample remained consistent. Later, the process by which sulfur was added to rubber to maintain its consistency became known as "vulcanization."

As metallurgy (the science of metals) developed, so did bicycle design. The key was to make the metal, and later composites of different types of metal, strong and light enough to support a rider without leaving him or her tired out by the effort expended in riding the bicycle.

In the late 1800s, bicycle tires were still made of solid rubber, but one day in 1887 John Dunlop's four-year-old son complained that his tricycle was too bumpy. Dunlop's answer? He made a "pneumatic" or air-filled tire for his son's tricycle. It worked. Suddenly, his son was riding on a cushion of air located between the tire and the rim. This part of the tire absorbed the jolt instead of the rider.

The next year he took out his own patent for a tough outer casing containing a soft, inflatable inner tube. By 1890, the pneumatic tire was in full-scale production, and it not only made bicycles more fashionable, but it also resulted in the design of bicycles with two equal-size wheels—the design that continues to be popular today.

The pneumatic tire was also quickly adopted to the wheels of cars, which in turn gave rise to the need for improved roads and driving surfaces. In fact, the age of the automobile owes as much to Dunlop's invention as it does to any other factor.

While the bicycle was important because it triggered other inventions, it is important in itself, being a key mode of transportation in many countries. Even in America, many people use it as their main mode of transportation, and in our health-conscious society it is a considerable asset.

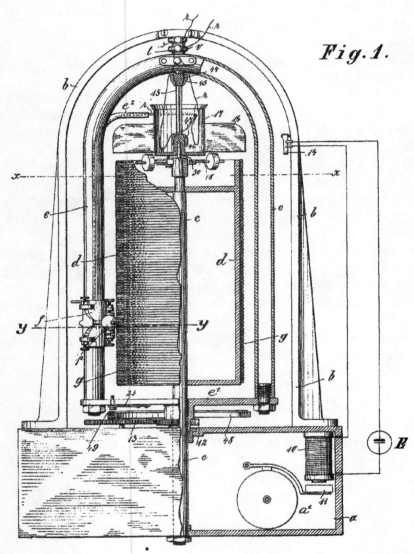

Patent drawing, 1900, by Valdemar Poulsen. *U.S. Patent Office*

Tape Recorder   96

During World War II, many people were surprised to hear German propaganda radio broadcasts late at night and early in the morning that featured symphony orchestras coming through the airwaves with remarkable fidelity, as if they were performing "live." U.S. Signal Corps technicians suspected that an improved recording device was being used, and after hostilities ended in 1945, such machines were discovered and brought to America for analysis. What the Germans had been using was an improved version of the "magnetophone" that was capable of capturing a frequency response of up to 10,000 hertz (cycles per second) with negligible noise and distortion.

The magnetophone patents came under the jurisdiction of the U.S. Alien Property office, but anyone could easily obtain a license through this office to develop his or her own machine. However, few decided to do so. The wire recorder was undergoing a resurgence in postwar America. With an improved response equivalent to that of the magnetophone and other "bugaboos" engineered out, firms such as Sears-Roebuck and Webster-Chicago (Webcor) aggressively marketed their products. The anticipated wire recording rage did not take place, however.

By 1946, the tape recorder was becoming the subject of increased interest. At this time, Bing Crosby, the famed "crooner," became important in tape recording progress. Crosby was the star of a popular radio show that showcased not only his singing, but also jokes and routines with guest celebrities. Crosby, a perfectionist, liked to prerecord the musical selections at a separate time and intermix them with other segments of the program. These in turn were rerecorded, by use of transcription disks, onto a "master" disk used for eventual airing. Unfortunately, transcription disks, though fine for original recordings, greatly degraded in noise and fidelity when multiple copies were made. When Crosby heard a demonstration of the captured German recorders, he quickly decided that tape was just the thing he needed. It could be easily cut, spliced, and rerecorded if a flub was made. With Crosby setting the pace, many other radio personalities, such as Jack Benny and Groucho Marx, started using tape recording on their popular broadcasts.

Three independent manufacturers—Magnecord, Rangertone, and Ampex—began manufacturing their versions of the German units for professional use in the mid-1940s. By 1950, they were in widespread use in recording, radio, and motion picture studios, replacing the disk and optical recorders then in use.

The Minnesota Mining and Manufacturing Company (3M) took up the task to develop better recording tape formulas. It created a new magnetic oxide that increased output and sensitivity. The new tape also featured an even coating for uniformity in recording and a plastic acetate base, which was to become the industry standard for over fifty years. The new plastic tapes quickly replaced the German paper ones—now a patchwork of splices and tears in the hands of the early broadcasters and experimenters.

Tape speeds could now be lowered as well. Speed past the recording head is a determining factor in reproduction of recorded audio. Human hearing detects frequencies between the range of about 30 to 15,000 cycles, so "high-fidelity" response, especially in music, must include this spectrum. It was eventually found that 1 inch per second (ips) of tape speed could recover about 1,000 cycles of sound. Therefore, operating at 30 ips, a tape recorder could achieve response to 30,000 cycles. The thirty ips speed was used until recently for critical classical recording, but for most purposes 15 ips became standard. This was later halved to 7½ and even 3¾, 1⅞, and ¹⁵⁄₁₆ ips for nonprofessional and special applications. With gradual improvements in both tape and recording technology, high-fidelity response became available at lower speeds by the 1970s.

Tape recorders were soon being marketed for consumer use by the late 1940s. The Revere Company promoted use of the 3¾ speed, which gave greater economy with the then expensive audio tape. Professional machines recorded across the whole width of the tape in one direction only. This provided fullest response and ease in editing. For home use, manufacturers used half the width of the tape, so that it could be turned over and used in the opposite direction. A 7-inch reel holding 1,200 feet of tape could provide up to 2 hours of recording time at 3¾ ips.

In the professional field, recording by tape became universal by the early 1950s. Motion pictures took advantage of the media to create the first "stereo" sound tracks, which are now being sold for the popular home video and digital video disk (DVD) market. Though the original sounds were rerecorded onto optical tracks for release—in monaural sound—to theatres, they can now be remixed and reprocessed with amazing results.

No less amazing was the evolvement of "multitrack" recording in the popular music field. In the late 1940s, guitarist Les Paul and his wife, singer Mary Ford, were frequently on the *Hit Parade* with their novel recordings that were dominated by multiple voices and guitar riffs. Paul had been producing these records in his home studio using multiple disks that he speeded up and dubbed to create his unique sounds. He quickly adapted tape to his production techniques and pioneered the tape "looping" and multitracking effects that were to create a sensation when used by artists such as the Beatles and the Beach Boys in the 1960s. He also helped design and develop the first multiple recording heads, which evolved into not only 2-channel stereo, but also 4-, then 8-, 16-, 24,- and 48-track recorders!

The perfection and marketing of stereo in the late 1950s, led to the evolution of home tape recording as well.

Simplicity of use and portability was the aim of two tape developments of the 1960s: the 8-track and the "cassette." The 8-track was an offshoot of a continuous-loop cartridge introduced for the broadcasting industry. This "cart," as it was called by studio technicians, was a single-track tape loop in a cartridge that ran at 7½ ips for up to 30 minutes. Most carts were built for up to 3 minutes of use in airing commercial announcements, station breaks, and hit recordings. The automobile industry and electronic firms had long been looking into producing a play-back system for cars. Records and reel-to-reel tapes had proven either unsuccessful or inadequate: Records skipped easily when the car hit a bump and drivers did not want to be distracted by threading or rewinding reels of tape. The Muntz Corporation and Lear Jet designers adapted the broadcasting cart for consumer use. Their 4-track (Muntz) and 8-track (Lear) slowed tape speed to 3¾ ips and used a moving head assembly to reposition the tape for succeeding stereo tracks. The 4-by-5-by-¼-inch cartridges were soon to be seen everywhere: not only in cars, but also in boats and planes and in nonportable versions in homes and restaurants. Thousands of long-play tracks were released in prerecorded 8-track versions. Prerecorded reel-to-reel tape had been around since the mid-1950s, but had not caught on as expected. Though 8-track and the less popular 4-track could provide good sound, they were soon plagued with technical problems not originally anticipated, so they quickly faded from sight—and sound.

The cassette—introduced by Norelco-Phillips in the early 1960s—was quite a different matter. It utilizes two reels of ⅛-inch tape housed in a compact 3⅞- by 2½-inch plastic cassette. Tape speed was slowed to 1⅞ ips to allow 60 minutes of recording, with the tape being flipped over after the first half hour. Thinner tapes allowed 90 minutes and eventually 120 minutes of playing and recording time. Though not as good in quality as the reel-to-reel recorder when introduced, now it produces remarkable recordings, especially when coupled to an external sound system.

97

From *Oklahoma Crude* (1973). *Photofest*

Oil Derrick

The first wooden structure that supported digging equipment over oil wells was named a "derrick." The word originally meant a "gallows," from a famous seventeenth-century English hangman named Derrick.

Like gallows, oil derricks used massive beams to support the weight of the drilling equipment. Unlike gallows, oil derricks are towers that taper at the top and use cross beams in X formations to support the structure.

The need for oil derricks closely follows the discovery and retrieval of oil from the ground. Although the combustible properties of oil had been known since ancient times, its collection before the development of derricks and drilling equipment was limited to sites where oil naturally and slowly percolated to the surface of the earth.

Oil had been used as a fuel for lanterns for many hundreds of years. Indeed, the Chinese had been drilling for it since the fourth century, but by the 1850s people were still skimming it off the surface of the earth. Inventors had realized that drilling for it was the best way to collect it, but no one could come up with the technology to do it.

Edward Drake changed this. In 1859, he built a derrick and a steam-powered drill and started probing the earth near Titusville, Pennsylvania. It was slow going, and a group of investors who backed Drake grew impatient. Indeed, so impatient that at one point they sent him a letter telling him to cease and desist. But in those days the mail was slow, and Drake kept going.

At one point, he had drilled down 69 feet and was about to stop working for the day when the drill dropped into an underground crevice. The next day, one of Drake's employees went out to check the drill rig. He looked down into the pipe that had been left in the hole. There, floating on top of the pipe, was oil. Drake had struck oil. A new industry was born.

Within a few years of Drake's discovery, America's oil industry was booming. The landscape became dotted with wooden oil derricks. Although at this time, and up until the turn of the twentieth century, some wells were still being drilled with a hand auger (as well as the steam engine), the oil derrick was really put to use when heavy drill bits were dropped into the ground to pulverize rock.

The pulverized rock was periodically bailed out and the process continued until the drill bit hit an oil deposit. Years later, cable-tool drilling rigs (designed to drill deeper) dotted the landscape. Heavy wooden derricks supported the equipment and provided leverage for pulling loose and heavy rock from the hole. The derricks supported tremendous weight as drilling bits and augers became heavier and cable was introduced to dig deeper.

The pressure at which the oil could be propelled from the earth could be immense, and the deposit of oil equally large. Once, in 1910, for example, a "blow out gusher" was hit and spewed out 9 million barrels of oil over 18 months!

Later, in the twentieth century derricks were built from metal and were often seen dotting the landscape with a "walking beam," the electrically driven arm that moves with a slow rocking motion. The beam is a plunger mechanism that pumps the oil out of the ground. The steel derricks with the rocker arms are often associated with the oil fields of Texas and other Western states.

No matter what type of derrick is being used, the principal is the same: Oil trapped in the ground needs to be brought to the surface for refining in the quickest and most cost-efficient manner possible. From the original wooden derricks that stood 80 feet tall, to the modern oil drilling rigs in the ocean, we have become quite successful at getting oil.

In the United States alone, the oil industry consists of some eight thousand companies and three hundred thousand workers, and there are oil reserves in more than thirty states. In a few states, such as Louisiana, Texas, Oklahoma, and California, millions of barrels of oil come out of the ground every day. But these reserves are slowing down to some degree.

Not to panic. There's still plenty of oil left. It's just a matter of getting the right technology to reach it. Drake went down 69 feet before he hit oil, but today there are wells that are 4,700 feet deep. The derrick is still a crucial aid in this deep drilling. Of course, it looks different in some cases from the original wood and metal derricks, but the purpose is still the same.

Ironically, oil fields that were abandoned as dried out have been given new life because of this technology. Today, derricks support cable drill bits that can not only go down thousands of feet directly, but also thousands of feet laterally. If oil is there, this technology has a good chance of finding it!

Thomas Edison's 1877 tinfoil phonograph. *U.S. Patent Office*

98

Phonograph

The phonograph is one of those inventions that is as romantic as it is practical. The invention occurred at a time in history that was rife with activity in electronics, acoustics (sound), and a philosophy that nothing was impossible. It also began a long line of similar inventions interrelated by sight and sound technology that were produced by different people in a relatively short period of time.

The creation of the sound production machines, including the phonograph, started as a way of producing historical archives. But as time went by, the direction shifted, and the phonograph soon emerged as a premier device for reproducing the sound of the singer and musical instrument.

It even reigned for several decades before the radio and sound motion pictures. The first successful sound recording device was developed by Leon Scott de Martinville in 1855. The device, named the "phonautograph," used a mouthpiece horn and a membrane linked to a stylus that recoded the sound waves on a smoke-blackened paper wrapped around a rotating cylinder. Beginning in 1859, the device was sold as an instrument for recording sound. But it had a major drawback: it could not play sounds back.

It wasn't until 1877 that Thomas Alva Edison designed the "tinfoil phonograph." Like many of Edison's inventions, his was the first to serve as a practical model that actually worked as he had intended it to.

For instance, Edison mixed creativity and practicality in his tinfoil phonograph by having the mouthpiece of the device replaced by a "reproducer," which was a more sensitive diaphragm. It had a cylindrical drum that was covered with tinfoil and mounted on a threaded axle. The device also had a mouthpiece that was connected to a stylus that etched the sound patterns on the rotating cylinder. The big plus of Edison's invention was that it could play back sound.

For the first demonstration, Edison said "Mary had a little lamb" into the mouthpiece. Although he was pleased that the experiment was successful, he was pleasantly shocked by the sound of his own voice, albeit tinny. Edison repeated the experiment for a friend of his at *Scientific American* magazine. The friend wrote this about the results of the experiment on November 17, 1877: "It has been said that Science is never sensational; that it is intellectual, not emotional; but certainly nothing that can be conceived would be more likely to create the profoundest of sensations, to arouse the liveliest of human emotions, than once more to hear the familiar voices of the dead. Yet science now announces that this is possible, and can be done. . . . Speech has become, as it were, immortal."

Edison got the credit for inventing the first "talking machine," perhaps in part because he was so famous and had unlimited money to produce prototypes and advertise himself, much like studios do for their actors when the Academy Awards roll around.

Still, he wasn't the first person to construct a phonograph machine. The first person to construct a workable design was a Parisian named Charles Cros. Producing plans that utilized disks, he presented his invention to the French Academie des Sciences in April 1877. To put this in perspective, this occurred several months before Edison stumbled onto the idea of a phonograph while working on a telegraphy device designed to record readable traces of a Morse code signal onto a disk.

All through 1878, Edison continued to fine-tune his phonograph, and the public never seemed to tire of him doing "performances" with it. He made a wide variety of sounds, from talking to coughing, and then, almost like magic, he reproduced the sounds. On a number of occasions, an audience member would try to prove it invalid by making a bizarre sound—such as the neigh of a horse—to see if the machine could reproduce it.

Unhappily for the phonograph's development, in 1878 Edison's fertile mind went elsewhere: to produce a working light bulb. Hence, he put aside the phonograph while he and his colleagues at his Menlo Park, New Jersey, laboratory focused on the light bulb, and nearly a decade went by before there was any further development on the device.

By the late 1870s and the early 1880s breakthroughs and rapid advances in communication technologies were giving the phonograph more attention. When

Edison renewed his interest in the phonograph, he insisted that it be used for more than entertainment.

The Bell-Tainter "gramophone" was released in 1887 and displayed some key improvements to his original model. By 1891, coin-operated phonographs were installed in drug stores and cafés that charged a nickel for approximately 2 minutes of music.

The commercial recording industry was said to have started by 1890. Musicians could record on several phonographs at once until enough cylinders were produced to meet demand.

Starting in 1901, Gramophone Company made sixty records by four stars of the Russian Imperial Opera. What followed was a world-wide industry that has continued until this day.

99

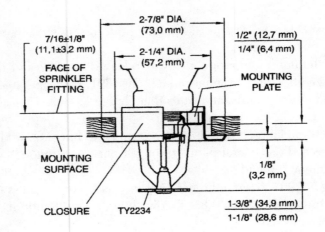

**FIGURE 2
STYLE 20 RECESSED ESCUTCHEON
FOR USE WITH THE SERIES LFII (TY2234)
RESIDENTIAL PENDENT SPRINKLER**

Diagram of residential pendent sprinkler. *Tyco*

Fire Sprinklers

The first fire sprinklers weren't concerned with protecting human life. Instead, they protected textile mills—their machinery and products—throughout New England. Nor were they automatic. If a fire started, the water was turned on so that it sprayed out through the holes in the perforated pipes. Inventors began experimenting with automatic systems around 1860. The first automatic sprinkler system was patented by Philip W. Pratt of Abington, Massachusetts, in 1872.

An American named Henry S. Parmalee is credited with inventing the first practical and fully operational sprinkler head in 1874 to protect his piano factory. From that time until the 1940s and 1950s, sprinklers were installed and used nearly exclusively in warehouses and factories, something motivated by insurance savings: building owners saved enough on reduced insurance premiums to pay for the cost of the sprinkler systems within a few years.

Eventually, fire sprinklers found their way into buildings where the primary purpose was to protect life. This development began after several fires that resulted in a large loss of life were investigated. Some of them included the Coconut Grove Nightclub in Boston in 1942 that killed a staggering 492 people (most of them by poisonous fumes from plastic accoutrements and by hazardous escape doors that were difficult to open), the Winecoff Hotel in Atlanta in 1946 that killed 119 people, and the LaSalle Hotel in Chicago in 1946 that killed 61 people.

Following these tragedies, investigators began noticing a pattern as they looked for ways to provide safety for building occupants: They found that factories, warehouses, and other buildings equipped with automatic sprinklers had amazingly good safety records compared to buildings that did not have sprinklers. As a result, authorities started requiring automatic sprinklers in certain buildings, particularly hospitals, government facilities, and other public buildings. They were (and still are) particularly valued in high-rise buildings where the sprinkler system was the only effective means of putting the fire out.

Today, fire sprinkler systems are made up of individual fire sprinkler heads and tubes that connect them. Usually, individual sprinklers are spaced throughout the ceiling of a building, then linked up with a network of piping and connected to a water supply. Heat from any fire activates one or usually several of the sprinklers in the immediate vicinity of the fire, allowing water to spray there, but not on the entire area that is protected.

To look at this a little more closely, when the heat of a fire increases a solder link within the sprinkler head will melt (at around 165 degrees F) or, depending on the design of the sprinkler system, a liquid-filled glass bulb will shatter to open the sprinkler head, releasing water directly onto the fire.

An important reason why fire sprinklers have such a good safety record is that they do not rely on human factors, such as familiarity with escape routes or emergency assistance, to operate. Instead, they are dormant and virtually maintenance free until they are needed. When needed, they go into action immediately. Characteristically, sprinklers prevent or retard fast-developing fires of intense heat, which are capable of trapping and killing building occupants in a very short period of time.

Some common potential problems people have with sprinklers are that they may go off in a fire and cause more damage than the fire itself. It has been found that sprinkler systems will cause far less damage than the smoke and fire damage if a fire goes unabated for some time. (Consider: "quick repose" sprinkler heads release 13 to 24 gallons of water per minute compared to 125 gallons per minute released by a fire hose.)

Modern sprinkler systems are continually and regularly tested to prevent accidental discharge and to make sure they are operable should a fire break out. In addition, sprinkler systems are designed for specific buildings and due to advances in sprinkler technology, unobtrusive and attractive ceiling- and sidewall-mount sprinkler heads can now be designed to match or blend in with room décor, while still being capable of providing effective fire protection.

Like plumbing systems, sprinkler piping is usually installed inside walls. This is done for a few reasons. One is that occupants do not have to look at piping inside the room. Another is that having the pipes inside the building eliminates their exposure to extreme weather temperatures.

Some interesting facts about sprinkler systems are that they rarely leak and are tested at 175 pounds per square inch (psi); standard plumbing is usually tested at 60 psi. Only the sprinkler head that is affected by the fire will activate. The heads are activated by heat, rather than by smoke. Sprinklers have been used since the late 1800s and are a proven, reliable, and safe technology.

100

Video recorder/player with camera. *Author*

Video Recorder

"Never miss your favorite television shows!" was the sales pitch that was used to promote the first home video recorders. Soon, just about everyone was getting to see those "late-show" movies that previously they had missed or had to stay up until 3 A.M. to watch.

The need to make a permanent record of television (TV) broadcasts became evident following the post–World War II development of modern TV. When network broadcasting became a reality in the late 1940s, the time difference between the East and West Coasts necessitated airing programs at unpopular times, restaging them "live," or recording them. All three options were utilized, although not with the videotape, of course.

The recording of video signals began as early as the 1930s, when Scottish pioneer John Logie Baird experimented with putting his photomechanical images on 78-rounds-per-minute (rpm) disks. This, however, proved unsuccessful.

Motion picture technology was soon put to use in creating the first practical—though not entirely satisfactory—method of TV recording. The "kinescope" was a specially designed motion picture camera that photographed the TV picture directly off the receiving monitor screen. Most kinescopes were made with 16-millimeter cameras, but some used the 35-millimeter format for higher quality.

There were a series of developments of the video recorder, but it was really the Sony company that got what was really home video recording off the drawing board and into our living rooms. Its "Betamax" was put on sale in 1975. The "Beta" system, which later proved to be a major disappointment, encased about 500 feet of ½-inch tape in a 6⅛- by 3⅝-inch plastic "cassette." The cassette was pushed down through a carriage on top of the deck and automatically loaded and threaded. The helical scanning tape heads were rotated at 1,800 rpm and resulted in a "writing" speed of 274 inches per second (ips), while the actual tape speed was 1.57 ips. This allowed a one-hour recording and playback time.

The Betamax produced high-quality full-color recordings, often indistinguishable from "live" reception. Though the first Betas retailed for $2,300, discounts were often available, and since there was no real alternative on the market, they sold well. A "battle of the formats" was on the horizon, however.

It quickly became apparent that the Beta's one-hour recording capability would not satisfy the home consumer. Taping movies became a major pastime of videocassette recorder (VCR) owners, and most of these ran 1½ to 2 hours or more. It was an inconvenience to change tapes, especially when using the built-in timer to capture the "late show" at 3 A.M. Furthermore, the early tapes retailed at $20 or more, which meant that a week's recording activity could run up a large price tag.

Japanese Victor Corporation (JVC), one of Sony's major competitors, wasted no time in taking advantage of the Betamax's only real fault. In 1976, JVC uncorked its Video Home System (VHS), which became the primary video recording format for the remainder of the VCR's existence.

JVC used the same 1/2-inch tape as Beta, but it was put into larger 7⅜- by 4-inch housing. Tape speed made the most important difference. VHS offered a choice of three: standard play (SP), operating at 1.31 ips, gave a 2-hour capacity; long play (LP), moving at 0.656 ips, allowed 4-hour continuous taping; and extended play (EP), creeping along at 0.437 ips, permitted 6 hours to be crammed onto a single cassette! The battle was on. Sony and its adherents correctly pointed out that the slowing of the recording speed resulted in both inferior picture and sound. But those inequities were negligible at the SP setting, and at least tolerable at LP and EP. Also, the added convenience and lower cost were just too tempting for the home user.

Sony responded to the challenge of VHS. It did so in the only way that was possible within the parameters of the Beta design—it simply slowed down the tape speed. "Beta I," as it came to be known, operated at about 1½ ips. Sony dropped it in favor of "Beta II," which ran at 0.787 ips, right in between VHS's SP and LP. This gave a 2-hour recording time, but now the famed Beta superiority in quality was compromised, as its picture and sound were no better than VHS. Sony eventually squeezed more tape into its cassettes and created a still slower speed, "Beta III," which accommodated 3 hours of recording time. The "extended" Beta cassettes eventually allowed 3.3 and 5 hours at the 2 speed options.

It proved "too little, too late," however. VHS's rising popularity soon saw not only JVC, but Panasonic, Radio Corporation of America, Sharp, and other brands taking over the market.

By the late 1970s, enterprising business firms obtained licenses to copy and distribute classic movies and music videos on both VHS and Beta. They originally retailed for anywhere between $29.95 and $79.95. Chains and individual proprietors soon sprung up to both sell and rent them. But prices eventually came down.

In its most popular form, the VCR consisted of a "deck" about 18 to 20 inches wide, 12 to 15 inches deep, and 5 to 7 inches high. The Beta was somewhat larger. Both weighed about 25 pounds. These dimensions shrunk drastically as circuitry became more compact and mechanisms were simplified. Many say the quality was reduced as well.

The tape was loaded in a well at the top of the deck, but this function was, by the early 1980s, moved to an autoloading slot on the face of the machine. "Piano" keys and later push-button relays controlled the standard tape functions of play, record, stop, fast forward, rewind, and pause—a carryover from home audio recorders. Remote controls, first attached by wire, then "wireless," became popular to control basic operation. The fast forward and rewind functions eventually included a "scan" feature that enabled viewing the picture while it was being shuttled back and forth, a big help in finding segments on a 6-hour tape.

Provision was made in the back of the deck for attachment of the TV antenna and cable to the receiver or monitor. Separate audio outputs were available. When stereo recorders were developed, they could feed exceptional sound—better than home reel-to-reel or cassette and close to compact disk (CD)—to outboard amplification and speaker systems.

Portable VHS and Beta recorders, later with built-in cameras, came on the market in 1978. They displaced home "super-8" moviemaking quite rapidly. By the late 1980s, home "camcorders" featured full-color and audio mixing and dubbing capabilities.

The 1980s and 1990s saw an incursion on the VCR market by manufacturers located in Korea, China, Malaysia, and other faraway places in the Orient and elsewhere. Brands such as Goldstar and Daewoo have taken the VCR prices down to a level unthinkable in the 1970s. The VCR, which had been expensive to both buy and repair, can now be had for under $100. Many "off-brands," "factory seconds," and "discontinued models" sell for around $50. Quality can be questioned, but the fact that the VCR became a part of everyone's life cannot be denied. Though the new digital video disks (DVDs) are rapidly making the VCR obsolete, many units—some that were bought over twenty years ago—are still working. When the VCR finally fades into the past, we'll all miss it, but we will have all the shows and events that we didn't miss because of it.

ACKNOWLEDGMENTS

I don't think a book of this size and scope can be done alone, and I was very fortunate to have some great people helping me, not only to research the entries but to write them.

Following are the people I want to thank profusely, and a little on their backgrounds. First, the writer/researchers:

Joe Beck. Joe teaches English in John F. Kennedy High School in Plainview, New York, as well composition at Kingsborough College. He has had a number of plays produced off-off Broadway, and has written for a variety of magazines. He is currently at work on his first book.

Joan Seaman. Joan is a graduate of Hunter College in New York City and has helped me write and edit a number of books, as well as research others.

Jessie Corbeau. Jessie is a writer and researcher and a very creative individual who is a silver jewelry designer as well.

Ruth-Claire Weintraub, MSW. Ruth-Claire, or Claire as I call her, said she fell in love with research when she was in college. I'm glad she did. I don't know where she learned to write, but she sure did.

Jules Rubenstein. Jules is a photojournalist, mainly working in the Bronx for local papers and has also worked for a wide variety of magazines. He is a technophile and sports buff and as knowledgeable as anyone I've ever met in these areas.

I also wish to thank my daughter-in-law Christina Philbin who gathered most of the illustrations for the book, making many trips to the New York Public Library and my son, Tom Philbin III, Chris's husband and my son, who is a professional photographer and provided a number of the photos for the book.

Thanks one and all!

INDEX

AC (alternating current), 161–62, 219; induction motor, 219–21
Aiken, Howard H., 24
Air conditioning, 141–43
Aircraft engine, 92–94
Airplane, 41–44
Alarm clock, 128
Alen, William van, 121–22
Ammann, Othmar, 146
Amperage, 255
Andrus, Leonard, 259
Anesthesia, 98–100
Apple Computer, 23
Arch bridge, 144–45
Arkwright, Richard, 195
Armour, Phillip Danforth, 266
Arrhythmia, 175–76
Arrowheads, 84–85
Atomic bomb, 53, 54–56, 229
Atomic reactor, 51–53
AT&T, 77–78
Automobile, 38–40

Babbage, Charles, 24, 225
Bachelor, Charles, 219
Bacon, Francis, 11
Bacon, Roger, 48–49
Bain, Alexander, 236–37
Baldwin, Frank, 225
Balloon framing, 208–9
Ballpoint pen, 34
Barbed wire, 163–65
Barbier, Charles, 137–38
Bardeen, John, 76, 77–78
Barker, George, 5
Battery, 101–3, 228–29
Beau de Rochas, Alphonse, 30–31

Beck, Vic, 184
Becquerel, Alexandre-Edmond, 236
Beehive oven, 269–70
Bell, Alexander Graham, 10–12
Bell Labs, 77, 78, 88, 231, 232–33
Bernstein, Julius, 173
Bessemer, Henry, 71, 72, 73
Betamax, 288–89
Bicycle, 271–73
Bifocals, 49–50
Binaural, 119
Birth control, 167. *See also* Condom
Black powder, 203. *See also* Gunpowder
Blackstone, William, 256
Blakewell, Frederick, 237
Blast furnace, 71–73
Blasting cap, 202–3
Bohr, Niels, 55
Booth, Major Edgar, 176
Boulton, Matthew, 81
Bow and arrow, 84–85
Brady, Matthew, 182–83
Braille, 137–38
Braille, Louis, 137–38
Brattain, Walter H., 76, 77–78
Braun, Karl, 18
Braun, Wernher von, 242–43
Bread baking, oven and, 268–70
Brick, 196–98, 270
Brooklyn Bridge (New York), 144, 146
Budin, Pierre, 151–52

Burroughs, William Seward, 225–27
Bushnell, David, 249
Bush, Vannevar, 23–24

Calculator, 225–27
Calley, John, 79, 81
Camera, 180–83; motion picture, 199–201
Cannon, 205–7
Carbon dioxide extinguisher, 263–64
Cardiac pacemakers, 175–77
Carré, Ferdinand, 266–67
Carrier, Willis, 141–43
Caselli, Giovanni, 237
Cassette tape recorder, 277
CAT scan. *See* CT scan
Cayetty, Joseph, 61
Cayley, George, 41, 43
Celsius, Anders, 148–49
Chain, Hans von, 93–94
Chariots, 1–2
Chronometer, 130–33
Chrysler Building (New York), 121–22
Churchill, Winston, 58, 139, 239
Circuit breaker, 254–55
Clock, 127–29
Cocaine, 99–100
Coffin, C. L., 87–88
Cohn, Alfred, 173–74
Colossus computer, 57–59
Colt, Samuel, 65–67
Columbus, Christopher, 83, 115, 131
Combustion engine, internal, 28–31

Index

Compass, 113–14
Computer: colossus, 57–59; desktop, 23–25
Condom, 166–68
Contact lenses, 50
Cook, James, 132
Cormack, Allan, 153
Cotton gin, 244–45
Cramer, Stuart, 143
Crompton, Samuel, 195
Crookes, William, 110–11
Crosby, Bing, 275
Cros, Charles, 282
CT (computed tomography) scan, 153–55
Cullen, William, 266

Daguerre, Louis-Jacques-Mandé, 181–82, 190
Daguerreotypes, 181–82, 190
Davenport, Thomas, 160, 161–62
Da Vinci, Leonardo, 26, 41, 223
Davy, Humphry, 98–99, 190
DC (direct current), 161–62, 219, 221
Deere, John, 31, 259
De Forest, Lee, 15, 77–78, 216–18
Densmore, James, 212
Desktop computer, 23–25
Dialysis machine, 178–79
Dickson, W.K. Laurie, 199, 201
Diesel engine, 31, 213–15
Diesel, Rudolph, 31, 213–15
Digges, Leonard, 170
Digges, Thomas, 170
Doge, John, 187
Drake, Edward, 278–79
Drywall, 158–59
Du Bois-Reymond, Emil, 173
Dunlop, John, 273
Duryea, Charles, 31
Dynamite, 202–4

Eastman, George, 180, 183, 191
Edison, Thomas Alva, 40, 219, 221; light bulb, 4–7, 217, 218, 229; motion picture camera, 199–201; phonograph, 281–83
Edoux, Leon, 125
8-track tape recorder, 277
Einstein, Albert, 51, 53
Einthoven, Willem, 173

EKG (electrocardiogram) machine, 172–74
Electrical wire, 75
Electric bulb, 4–7
Electric elevator, 125–26
Electric motor, 160–62
Electromagnetic induction, 161–62
Electromagnetic radiation, 190–91
Electrons, 218
Elevator, 124–26
Empire State Building (New York), 121–22
Engines: aircraft, 92–94; diesel, 31, 213–15; internal combustion, 28–31; jet, 92–94; rocket, 240–43; steam, 28, 41, 43, 79–81, 95
Enigma code, 57–59
Erickson, Leif, 115
Ether, 99
Evans, Oliver, 266
Eveready, 229
Eyeglasses, 48–50, 169–70

Fahrenheit, Daniel Gabriel, 147–48
Faraday, Michael, 11, 13, 98–99, 160–61
Farnsworth, Philo T., 15
Fax (facsimile) machine, 236–37
Fermi, Enrico, 51–53
Fibrillation, 175–76
Film, 182, 183, 190–92
Finley, James, 145
Fire extinguisher, 262–64
Fire lance, 206
Fire sprinklers, 284–86
Fireworks, 20, 22, 241
Fisher, Alva J., 257
Flashlight, 228–29
Fleming, John, 15, 218
Flintlock, 63, 65
Flying shuttle, 194, 195
Ford, Henry, 31, 39–40
Forsyth, Alexander John, 63
Fountain pen, 32–34
Framing, balloon, 208–9
Franklin, Benjamin, 33, 49–50, 69
Freidel, Robert, 7
Freon, 267
Freud, Sigmund, 100
Fuare, Emile Alphonse, 102
Fulton, Robert, 234–35
Fuse (circuit breaker), 254–55

Galileo, 128, 134, 135, 147, 169, 170
Galvani, Luigi, 101
Gas engine, 31, 213, 215
Gassner, Carl, 228
Gas welding, 87–88
Gates, Bill, 23
Gates, John, 165
General Electric (GE), 257
George Washington Bridge (New York), 144, 146
Glidden, Joseph Farwell, 163, 164, 165
Global positioning system (GPS), 184–86
Goddard, Robert Hutchings, 240–43
Goldsmith, Michael, 156
Goodyear, Charles, 166, 273
Gordon, James P., 232
Gorrie, John, 141, 266
Gothic typeface, 9
Gramophone, 283
Gray, Elisha, 10–11
Great Train Robbery (movie), 201
Great Wall of China, 198
Guggenheim, Daniel, 242
Gunpowder, 20–22; and cannon, 205–7
Gutenberg, Johannes, 8–9, 36
Gypsum, 158–59

Halley, Edmond, 131–32
Hargreaves, James, 193, 194, 195
Harrington, John, 61
Harriot, Thomas, 170
Harrison, John, 131–32
Helicopter, 222–24
Hemodialysis, 178–79
Hemp, 82–83
Henry Ford Company, 39–40
Henry, Joseph, 26, 27, 161, 255
Hertz, Heinrich, 17
Hindle, Charles, 173
Hiroshima (Japan), 54–56
Hobnail, 104
Holland, John P., 249–51
Hounsfield, Godfrey, 153
Howe, Elias, 188, 189
Hubble Space Telescope, 170–71
Hulney, 249
Hunt, Walter, 187
Hussey, Obed, 90
Hyman, Albert S., 176

Index

IBM, 24, 25
Iconoscope, 14, 15–16
Incandescent light bulb, 4–7
Incubator, 150–52
Internal combustion engine, 28–31
Iron-to-steel process, 71–73, 104–5

Jackson, Charles T., 99
Jansen, Zacharias, 134–35
Jenkins, Charles Francis, 13
Jenney, William Le Baron, 121
Jet engine, 92–94
Jobs, Steve, 23
Josephson, Matthew, 6
Joule, James, 255
Julian, Percy Lavon, 263

Kane, Fred L., 159
Kay, John, 194, 195
Kelly, William, 71, 73
Kidney dialysis machine, 178–79
Kiln, 197
Kinescope, 287–88
Knight, Godwin, 114
Knowles, John, 187
Kodak, 183, 191
Krayer, Otto, 176

Laënnec, René, 118–19
Laser, 230–33
Latex paint, 253
Latitude, 130–33
Laughing gas, 98–99
Leclanché, Georges, 228
Lenard, Phillipp, 110–11
Lenoir, Etienne, 30–31
Lenses, 49, 50; and telescope, 169–70
Lidwill, Mark C., 176
Light bulb, 4–7, 229
Lilienthal, Ott, 43
Lindbergh, Charles, 242
Lippman, Gabriel, 173
Livingston, Robert, 235
Locomotive, 95–97, 145
Longitude, 130–33
Los Alamos (New Mexico), 53, 55–56
Lumière, Auguste, 191
Lumière, Louis, 191, 201
Lundy, John, 100

McCormick, Cyrus, 89–91
Magnetic compass, 113–14
Magnetophone, 275

Manhattan Project, 53, 55–56
Marconi, Guglielmo, 15, 17–19
Martin, Odile, 150, 151
Martin, Thomas J., 263
Meikle, Andrew, 259
Mercury thermometer, 147–49
Metcalf, Thomas B., 227
Meteorology, 140
Microscope, 134–36
Microsoft, 23
Microwave spectroscopy, 231–33
Minié, Claude-Étienne, 63–64
Minkoff, Larry, 156
Misell, David, 229
Model T, 40, 109
Moon, William, 138
Morphine, 100
Morse, Samuel F. B., 26–27
Morton, William Thomas Green, 99
Motion picture camera, 199–201
MRI (magnetic resonance imaging), 156–57
Musket, 63
Muybridge, Eadweard, 199, 201

Nail, 104–6, 208, 209
Napier, John, 225
Newcomen, Thomas, 79–81
Newton, Isaac, 93, 131, 170
Niemann, Albert, 99
Niepce, Joseph-Nicéphore, 181–82, 190
Nipkow, Paul, 13
Nitroglycerine, 202–3
Nitrous oxide, 98
Nobel, Alfred, 202–3
Nuclear bomb, 53, 54–56, 229
Nuclear energy, 51–53

Oil derrick, 278–80
Oppenheimer, J. Robert, 53
Otis, Elisha Graves, 124, 125, 126
Otto, Nikolaus, 28, 30–31
Oven, 268–70

Pacemaker, 175–77
Pacemaker cells, 172–73
Paint, 252–53
Pantelegraph, 237
Paper, 35–37
Parmalee, Henry S., 284

Pascal, Blaise, 225
Paul, Les, 276–77
PC (personal computer), 23–25
Pendulum clock, 128–29
Pen/pencil, 32–34
Peritoneal dialysis, 178–79
Phillips head screw, 108
Phonograph, 281–83
Pigments, 252–53
Piorry, Pierre Adolphe, 119
Pistol, 65–67
Plante, Gaston, 102
Plaster (plastering), 158–59
Pleximeter, 119
Plow, 45–47
Plumbing, 68–70
Plutonium, 55
Popov, Alexander Stepanovich, 17–18
Porter, Edwin, 201
Potassium nitrate, 205–6
Potter's wheel, 2–3
Pratt, Philip W., 284
Printing press, 8–9, 36

Quartz clock, 129
QWERTY keyboard, 212

Radar, 139–40
Radio, 15, 17–19
Radiography, 110–12
Reaper, McCormick, 89–91
Refrigerator, 265–67
Remington, Philo, 212
Richardson, Owen W., 218
Rifle, 62–64
Rittenhouse, William, 36
Robertson, Peter L., 108–9
Rocket, 240–43
Roebling, John A., 75, 145–46
Roebling, Washington, 146
Roman typeface, 9
Rontgen, William Conrad, 111–12
Roosevelt, Franklin D., 51, 53, 55, 167
Rubber, and bicycle, 271, 273
Rubber condoms, 166–68

Sackett, Augustine, 159
Sail, 82–83, 117, 246
Saltpeter, 205–6
Sanger, Margaret, 167
Sarnoff, David, 15–16
Saxony wheel, 193–94
Schawlow, Arthur L., 230–33
Screw, 107–9

Index

Senning, Ake, 177
Sewing machine, 187–89
Shockley, William, 77–78
Sholes, Christopher Latham, 210–12
Siemens, Werner von, 126
Sikorsky, Igor, 222–24
Simms, Frederick, 238–39
Singer, Isaac, 189
Skyscraper, 120–23
Slide rule, 225
Smith, Horace, 67
Smyth, E. W., 217
Socket head screw, 108–9
Sodium pentothal, 100
Sonar radar, 140
Sony, 288–89
Space flight, 241–43
Spectacles, 48–50, 169–70
Spinning jenny, 193–95
Spoke and wheel, 1–3
Steamboat, 234–35
Steam engine, 28, 41, 43, 79–81, 95
Steel. *See* Iron-to-steel process
Stephenson, George, 96
Stethoscope, 118–19
Stilwell, Charles, 37
Street, Robert, 29–30
Submarine, 249–51
Sullivan, Louis, 121
Suspension bridge, 75, 144–46
Szilard, Leo, 51–53

Talbot, William Henry Fox, 182–83, 191
Tank, 238–39
Tape recorder, 274–77

Tarnier, Stephane, 151
Telegraph, 26–27, 217
Telephone, 10–12
Telescope, 169–71
Television (TV), 13–16, 78, 287
Telford, Thomas, 145
Terra-cotta, 198
Tesla, Nikola, 161, 219–21
Thermometer, 147–49
Thompson, Elihu, 86, 88
Thompson, William, 102
Thomson, Joseph J., 218
Thread, 193–95
Threshing machine, 259–61
Titanic, 19
Toilet, 60–61
Townes, Charles H., 230–33
Transistor, 76–78
Trevithick, Richard, 95–96
Trilateration, 184–85
Triode vacuum tube, 77–78, 216–18
Truman, Harry S., 55
Tsiolkovsky, Konstantin, 241
Turing, Alan Mathison, 57–59
Twain, Mark, 212
Typewriter, 210–12

Vacuum tube, 77–78, 216–18
VD (venereal disease), and condoms, 166–67
Verne, Jules, 241
VHS (video home system), 288–89
Video recorder, 287–89
Viking ships, 115, 116–17
Visible spectrum, 190–92
Volta, Alessandro, 101–2

Waller, Augustus Desiré, 173
Washburn, Ichabod, 74–75
Washing machine, 256–58
Washington, George, 33
Waterman, Lewis, 32, 34
Waterwheel, 3, 247
Watson-Watt, Robert, 139–40
Watt, James, 81, 95
Wedgwood, Thomas, 182, 190
Weisenthal, Charles, 187
Welding machine, 86–88
Wesson, Daniel B., 67
Westinghouse, George, 221
Wheel, 1–3
Whiter, Abraham, 217–18
Whitney, Eli, 26, 195, 244–45
Whitney, Eli, Jr., 67
Whittle, Frank, 92–94
Wilson, Allen, 189
Windmill, 246–48
Wire, 74–75, 145–46; barbed, 163–65
Wooden ships, 115–17
World Trade Center (New York), 122–23
Wright, Frank Lloyd, 121
Wright, Orville and Wilbur, 42, 43–44

X-ray machine, 110–12, 153–54

Zeiger, H. L., 232
Zinn, Walter, 51
Zoll, Paul Maurice, 176
Zoopraxiscope, 199
Zworykin, Vladimir K., 14, 15–16